3-28-15

BUSINESS DIVISION

CHICAGO, IL

PRACTICAL BUILDING CONSERVATION
VOLUME 2
BRICK, TERRACOTTA AND EARTH

Practical Building Conservation Series:

Volume 1: STONE MASONRY
Volume 3: MORTARS, PLASTERS AND RENDERS
Volume 4: METALS
Volume 5: WOOD, GLASS AND RESINS

PRACTICAL BUILDING CONSERVATION

English Heritage Technical Handbook

VOLUME 2

BRICK, TERRACOTTA AND EARTH

John Ashurst
Nicola Ashurst

Photographs by Nicola Ashurst
Graphics by Iain McCaig

HALSTED PRESS
a division of JOHN WILEY & SONS, Inc.
605 Third Avenue, New York, N.Y. 10158
New York • Toronto

© English Heritage 1988

All rights reserved. No part of this publication may
be reproduced in any form without the prior permission of
John Wiley & Sons, Inc.

Published in the U.S.A. and Canada
by Halsted Press, a division of
John Wiley & Sons, Inc., New York

Library of Congress Cataloging in Publication Data available:

ISBN 0–470–21105–9

TH
3351
.A84
1988
v. 2

Printed in Great Britain

BUSINESS/SCIENCE/TECHNOLOGY DIVISION
CHICAGO PUBLIC LIBRARY
400 SOUTH STATE STREET
CHICAGO, IL 60605

CONTENTS

Foreword

by Peter Rumble CB, Chief Executive, English Heritage

Over many years the staff of the Research, Technical and Advisory Service of English Heritage have built up expertise in the theory and practice of conserving buildings and the materials used in buildings. Their knowledge and advice have been given mainly in respect of individual buildings or particular materials. The time has come to bring that advice together in order to make available practical information on the essential business of conserving buildings – and doing so properly. The advice relates to most materials and techniques used in traditional building construction as well as methods of repairing, preserving and maintaining our historic buildings with a minimum loss of original fabric.

Although the five volumes which are being published are not intended as specifications for remedial work, we hope that they will be used widely by those who write, read or use such specifications. We expect to revise and enlarge upon some of the information in subsequent editions as well as introducing new subjects. Although our concern is with the past, we are keenly aware that building conservation is a modern and advancing science to which we intend, with our colleagues at home and abroad, to continue to contribute.

The Practical Building Conservation Series

The contents of the five volumes reflect the principal requests for information which are made to the Research, Technical and Advisory Services of English Heritage (RTAS) in London.

RTAS does not work in isolation; it has regular contact with colleagues in Europe, the Americas and Australia, primarily through ICOMOS, ICCROM, and APT. Much of the information is of direct interest to building conservation practitioners in these continents as well as their British counterparts.

English Heritage

English Heritage, the Historic Buildings and Monuments Commission for England came into existence on 1st April 1984, set up by the Government but independent of it. Its duties cover the whole of England and relate to ancient monuments, historic buildings, conservation areas, historic gardens and archaeology. The commission consists of a Chairman and up to sixteen other members. Commissioners are appointed by the Secretary of State for the Environment and are chosen for their very wide range of relevant experience and expertise. The Commission is assisted in its works by committees of people with reputation, knowledge and experience in different spheres. Two of the most important committees relate to ancient monuments and to historic buildings respectively. These committees carry on the traditions of the Ancient Monuments Board and the Historic Buildings Council, two bodies whose work has gained them national and international reputations. Other advisory committees assist on matters such as historic gardens, education, interpretation, publication, marketing and trading and provide independent expert advice.

The Commission has a staff of over 1,000, most of whom had been serving in the Department of the Environment. They include archaeologists, architects, artists, conservators, craftsmen, draughtsmen, engineers, historians and scientists.

In short, the Commission is a body of highly skilled and dedicated people who are concerned with protecting and preserving the architectural and archaeological heritage of England, making it better known, more informative and more enjoyable to the public.

ACKNOWLEDGEMENTS

The authors gratefully acknowledge the assistance of Dr Clifford Price, Head of the Ancient Monuments Laboratory, English Heritage, in the reading of the texts.

1 CONTROL OF DAMP IN BUILDINGS

1.1 CAUSES OF DAMP

Dampness in historic masonry may be due to one or more of the following causes:

- Direct penetration of rainwater
- Rising damp (water from the ground conducted by capillarity)
- Hygroscopic (moisture-attracting) salts, usually introduced as the result of rising damp
- Condensation (water from air within the building)

It is important during the analysis of a damp problem that the exact cause or causes are identified. This is critical to the selection of an appropriate method of rectification, particularly as a solution for one cause may exacerbate a damp problem due to another cause. The correct diagnosis of the cause of dampness is greatly assisted by a map of the distribution of moisture within a wall and quantification of the degree of dampness. It should be remembered that moisture meters and other methods of determining the presence of moisture in building materials cannot differentiate between dampness from one source and that from another.

This chapter considers the four principle causes of damp and evaluates several types of treatment method.

1.2 PENETRATING DAMP

Walls, particularly those with defective pointing or cracked external rendering, may be penetrated by rainwater reaching them through exposure to the prevailing wind, or from defective rainwater goods and poor construction details. The dampness may affect all or part of the wall and is likely to be particularly apparent during or directly after periods of heavy rain. Damages and defects in roof coverings, gutter linings and flashings are another common contribution to penetrating damp. Where the internal floor level is below external ground level, moisture will penetrate laterally unless blocked by a vertical damp-proof membrane. In areas where the water table is high, lateral penetration of dampness may be aggravated by hydrostatic pressure. In these situations the dampness must be stopped at the point of penetration.

1.3 RISING DAMP

Characteristics

Rising damp is characterized by a descending moisture gradient within a wall from ground level sometimes to a height up to 1.5 metres (5 feet). Where an impermeable barrier lines the wall at low levels, the level can be even higher. This gradient will not be detected by electric moisture meters which operate on the wall surface, but will need to be identified from samples taken from within the wall. Rising damp is frequently accompanied by salt migration to the wall surface within the wetting and drying zone at the top of the damp area. The accumulative volume change associated with the crystallization of all such salts eventually causes serious breakdown of the masonry surface. Sulphates usually manifest themselves as efflorescence, while hygroscopic salts, chlorides and nitrates, may also appear as damp patches.

The movement of water

Water moves through a masonry wall by capillarity, the phenomenon whereby liquids are drawn into very fine tubes or pores within a material (capillaries). The water rises through the network of capillaries moving against the force of gravity and in all directions, to a level beyond its source. The properties of each material involved (in particular capillary size), as well as the thickness of the wall, affect the height of rise.

Evaporation occurs at the wall surface over the area of capillary rise and has the effect of restricting the height of the rising damp. A 'hygric' balance is achieved when the rate at which the wall absorbs water is balanced by the rate of evaporation. The horizontal section area of a wall at ground level directly affects the quantity of water taken up and this intake is balanced by the total evaporation area of the wall. Therefore, other conditions being equal, a thick wall with a large amount of absorption needs a large surface area for evaporation. In other words, as a general rule, the thicker the wall, the higher the damp will rise.

Presence of salts

At the top of a rising damp zone in a wall evaporation and dampness alternate and the masonry undergoes repeated cycles of wetting and drying. As this usually takes place in the presence of soluble salts which have migrated upwards from the ground, salt crystallization damage occurs. Below this 'front' the salts are mostly in solution and the damage is much less.

Amount of absorption

The absorption potential of a masonry wall is that of the predominant material. The thinner the joints and the less mortar used in the construction, the more a wall's behaviour resembles that of its stones or bricks. A structure containing impermeable material such as granite or vitrified brick is slow to accept dampness because the only route for the moisture is through the mortar which is relatively slow in absorption. In a more permeable material such as chalk or gauged brick, the masonry units can act as a shortcut between the mortar layers and the wall is invaded by dampness quite quickly.

The effect of waterproof surfaces

It is often incorrectly thought that the damage caused by rising damp can be stopped by eliminating evaporation. Waterproof surfaces such as impermeable cement-based renderings will only force dampness to rise above the surfacing to re-establish the necessary area of evaporation, thus extending the area affected.

Sources of water

The most common sources of water for rising damp are concentrated localized supplies such as ponding rainwater or leaks from supply or disposal pipes; all can produce areas of soaking of the ground in contact with the building footings. When the water table is the source of dampness, all the walls in contact with the ground are likely to be affected and all will receive an equal amount of water. The level of dampness on internal walls can be higher than that on external walls because of the reduced rate of evaporation. The distribution of dampness due to localized water sources will usually be well defined. Outside walls are sometimes affected by localized water sources originating from outside the building perimeter.

Masonry walls will always contain some water. Preliminary experiments conducted in Rome found 3 per cent for the brick tested and 5–6 per cent for the sandstone tested to be acceptable.

1.4 CONTROL OF RISING DAMP

The simplest and most effective way to eliminate rising damp is, when feasible, to remove the source of dampness. The control of localized water sources may involve improving the collection and removal of water from the roof, including attention to parapet gutters, flashings and hopper heads. Surface water on the ground should not be allowed to pond at the base of a wall and water supply and sewerage systems must be checked for damage and made fully operational. Significant drying can often be achieved by reducing the site water level with external land drainage.

There will, however, still be times when it is necessary to establish a barrier within the masonry to prevent water rising. No treatment of rising damp should be undertaken without first considering the likely effects of changing the moisture content within a wall. Consideration must be given to the removal of salts after the installation of a damp-proofing system. A damp-course installation coupled with the introduction of central heating will frequently bring massive efflorescence and disruption of internal plaster as the damp zone dries out. In spite of many advances in the treatment of rising damp, the only positive control is by the physical insertion of a suitable impervious barrier. Such insertions, however, are normally restricted to walls of regularly bonded construction and limited thickness. Historic buildings may require another method such as chemical injection or transfusion.

Drying aids

Air drains and dry areas

Early attempts to overcome damp in walls at ground level were the provision of air drains and dry areas, sometimes as part of the original construction, and sometimes introduced as a remedial measure. The principle involved was simply to isolate the occupation area from the wet ground with a retaining wall and a ventilated space, and to encourage maximum drying out of the wall below ground level. This is partially achieved in some cases by digging out dry areas at the base of the walls, laying land drains to falls in the base of the trench so formed, and backfilling with a large, dry ballast. The depth of excavation can be small, and in many cases should be restricted to 300–400 mm. Disturbance of this kind should not be carried out indiscriminately on sites of historical significance without archaeological advice and supervision. There may be other cases where shallow footings or difficult site conditions make this kind of operation impracticable and structurally inadvisable.

Other drying may be encouraged by the removal of dense impervious materials adjacent to the wall. Dense paving should not be butted to wall bases in dense mortar. A dry margin, even a small one of 50 mm, without an attempt to exclude water should be removed. Dense cement or hydraulic lime pointing should be cut out and substituted with a weaker cement/lime/sand mix.

High-capillary tubes

In the last century a perforated fireclay brick course was sometimes installed to encourage drying out, and of subsequent attempts to increase drying in the wall the 'high-capillary' tube system has been most extensively used. This system introduced small earthenware tubes into previously drilled holes in the wet wall up to a shallow angle of 10°–15°. The holes penetrate two-thirds of the wall thickness. The tubes are bedded into the holes in a mortar which should be as weak and laid as dry as possible.

The objective is to attract moisture to the drying tubes where it is encouraged to evaporate as quickly as possible. There is, however, an inherent problem here, since tubes which are capable of attracting water by capillarity will inevitably be reluctant to lose that water by evaporation. They can also become hygroscopic by the deposition of soluble salts. Of course the holes must be spaced fairly closely together and preferably installed in two or three staggered courses to have any significant drying effect, and in such circumstances they also function as a partial damp-proof course, since water cannot cross them. Installations of this type cannot be expected to solve serious damp problems, although they may sometimes have a useful role to play when there is penetrating damp or where they are

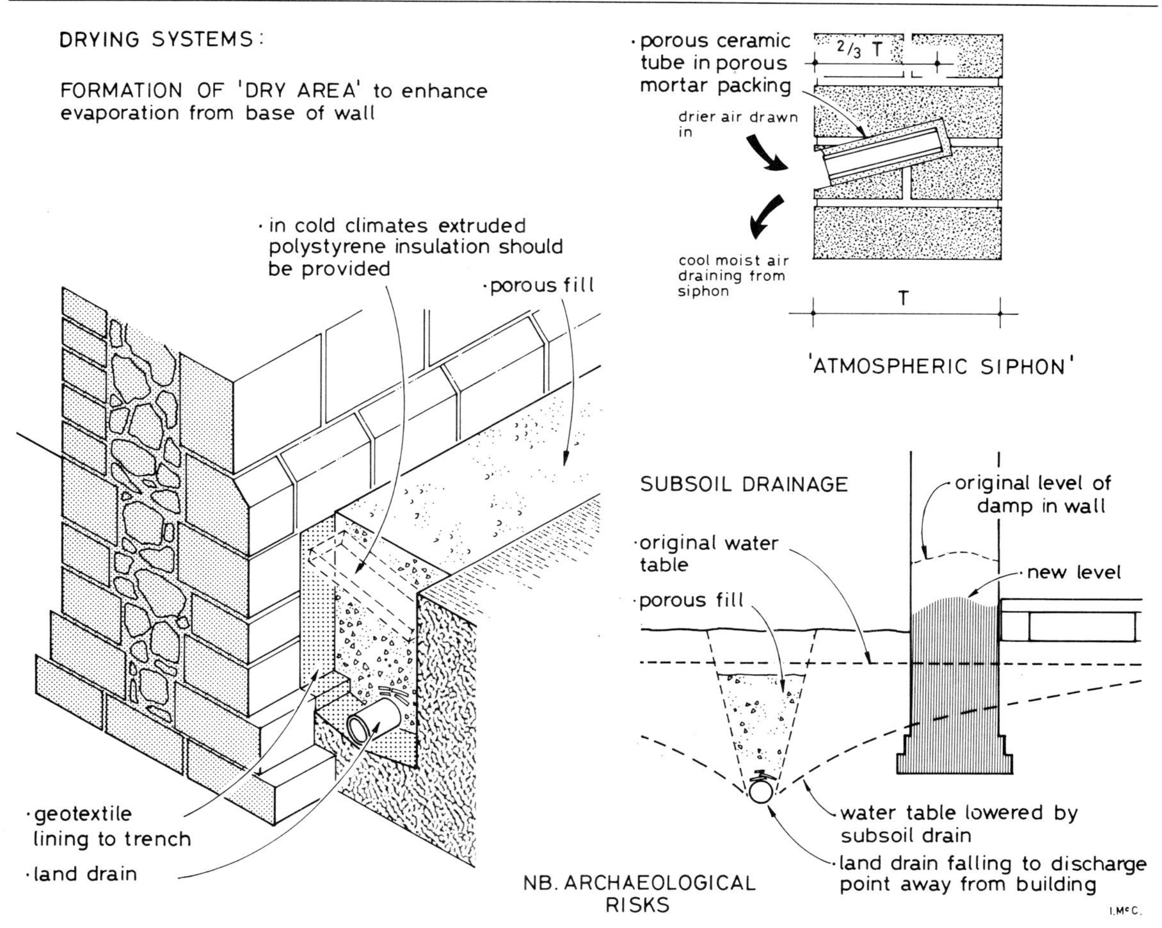

Figure 1.1 Drying aids

used in conjunction with another installation to check the amount of drying out that takes place on the internal wall surfaces. However, practical evidence has shown the high-capillary tube system to be of very limited effectiveness in the control of rising damp and in some circumstances the tubes could increase the humidity locally.

Useful commentary on high-capillary tubes can be found in the references at the end of this chapter under Torraca and Richardson.

Physical barriers

The horizontal cutting through a wall and the insertion of a positive damp-proof course is an effective system of preventing dampness due to capillarity.

The traditional method of inserting a damp course in brickwork is to cut out short lengths of brick course at a time and replace them with dense engineering bricks or slates. Less costly is cutting or grinding through a bed joint with a hand saw or power saw in lengths of about 0.5 m and inserting an impervious sheet membrane such as half hard copper.

Some of the most efficient cutting is now carried out with chain saws cutting a 7 mm wide slot. Chain saws are available to cut through up to 1.5 m thicknesses. Another useful cutting tool is a flexible glass fibre disc impregnated with carborundum up to 0.6 m diameter, driven by a portable motor with a flexible drive and fitted with a vacuum extract which eliminates almost entirely the enormous dust problem associated with this operation. *BRE Digest 27* (1970) 'Rising Damp in Walls' gives details of cutting procedures.

One of the major problems associated with cutting through a wall, however carefully it is packed and supported as work proceeds, is the risk of settlement after installation. Cutting through a wall also causes major disturbance to internal surfaces.

An Australian system which involves a series of heavy-duty polyethylene sheet envelopes seeks to resolve the settlement problem. The envelopes are slid into a wall as the saw is removed and are then filled with grout under pressure following the insertion and overlapping of the subsequent envelope.

Metal sheets with protective coatings have been used to form a positive damp-proof course. Experience has shown that coatings are often damaged in transit and during insertion, leaving the metal, commonly aluminium, victim to salt attack. High density, heavy duty polyethylene sheet, sometimes filled with carbon black, is likely to be more reliable than standard-gauge metal barriers.

A versatile physical insertion system is that developed by Doctors G and I Massari in Italy. The Massari method involves the insertion of polyester resin in a cut through the full thickness of a wall to form a damp-proof course which structurally integrates the wall above and below the cut. Walls of up to 1200 mm thick and of solid brick or stone construction are cut by chain saw; where this may cause unacceptable amounts of vibration in walls of a sensitive nature, varied construction and up to 3.5 mm thick, the cut is made using a diamond coring drill. A slot 420 mm long is formed by a series of 25 mm diameter drill holes; it is then thoroughly dried and a specially designed trough fixed to each side. Commercially available polyester resin is poured into these to a level above the top of the cut. After initial curing has taken place, the resin is trimmed off along the line of the wall. The complete barrier is formed in alternate sections so that no section of wall longer than 420 mm is ever left unsupported.

The Massari system has been used both horizontally and vertically to isolate valuable plaster and frescoes. Whilst knowledge of the lifetime of resins is still limited, a system of this kind is certainly more positive and will be longer lasting than, for instance, a single chemical injection with resin. It is, of course, proportionally expensive, and there may be other reasons for objection to the amount of drilling, even with low-vibration equipment.

Chemical systems

Some of the chemical damp-proof courses available are currently the most promising of commercial installations for very thick masonry walls. The principle is

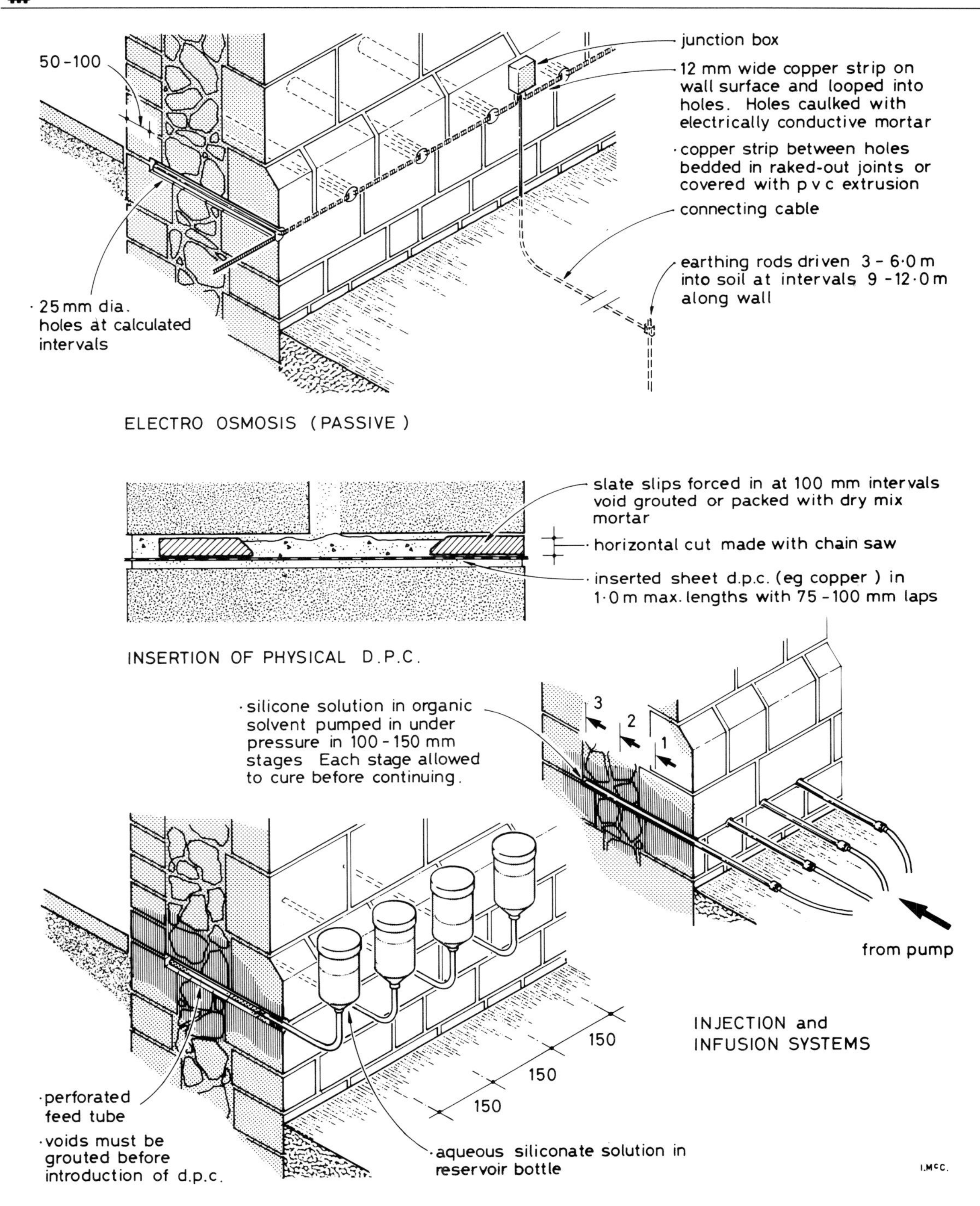

Figure 1.2 Damp-proofing systems

to establish a water-repellent zone across the wall in the area just above ground level where a damp-proof course would normally be installed.

Materials include aluminium stearates, silicone solutions in an organic solvent, and water-soluble silicone compositions (siliconates) that become water-repellent on curing. These materials line the walls of the pores of the masonry and change the contact angle of the meniscus, thus inhibiting the capillary attraction up the wall. One system adds a rubber latex emulsion intended to block the pores with a flexible material.

Aqueous siliconate solutions are usually allowed to drip feed into the wall from rows of reservoir bottles. The silicone solutions in solvent and the siliconate/latex mixture are normally pumped in under pressure. Both require a line of close-spaced drillings (average 150 mm centres) and both require experienced contractors to carry out the installation. The drillings are obviously best made in the thickness of mortar joints wherever possible, since filled drillings in masonry always show and tend to become more obtrusive with age. Success of this type of system depends on uniform diffusion through the thickness of the wall. In typical medieval construction with two skins of face work and a rubble fill it may well be necessary to carry out a grouting operation of the damp zone before spending money on introducing expensive chemicals in to fill a wall with a high percentage of void.

One system involves the placement of frozen pellets of siliconate solution into holes drilled into a wall. The pellets melt, diffuse into the wall and the solution cures to form a barrier. Another system involves injection of a special mortar into a similar series of holes. In the presence of water, deposition and spread of insoluble, pore-blocking salts from the mortar takes place. The main disadvantage of both these systems is that it is difficult to place enough material in the wall for them to be effective.

Professional advice is recommended before attempting the chemical injection of a wall which is extremely wet and likely to be contaminated with salt. In addition to other precautions, stone paving against a wall should be protected against excessive spillage of silicone, which may have the effect of forming a shallow surface skin vulnerable to spalling in the presence of moisture and salt.

Electro-osmotic systems

Various electrical systems are available which claim to inhibit the passage of water up a wall. Passive systems seek to 'earth' the wall by linking electrodes in the wall with electrodes of similar metal in the ground, or to produce an electrical potential by galvanic action using dissimilar metals for the two sets of electrodes. Active systems apply an electrical potential from an external source between the sets of electrodes. Only the active system can accurately be described as 'electro-osmotic', but no difference in terminology is made commercially.

The electrodes are set in holes equal to approximately half the depth of the wall + 5 cm. On a wall 1 metre thick the electrodes would need to be set in 500 mm and spaced approximately 650 mm apart. In a positive system, these electrodes would be linked to an 'earth stake', of, for example, 15 mm copper tube. An alternative arrangement places the electrodes in the wall at a depth of 150 mm in a regular triangular pattern with 1 metre spacing.

Unfortunately, although it has been recognized for over a hundred years that walls exhibiting symptoms of rising damp produce an electrical potential, and that the application of a DC voltage to linked electrodes buried in a wet wall and in the ground might control the passage of water, there is still a dearth of evidence that an 'electro-osmotic' system has been significantly successful. The reasons for this lack of evidence need not be that an active system does not work; completed installations are almost always carried out in conjunction with replastering and often with improved drainage and heating, operations which may make considerable difference to wet walls even without an installation.

One of the major problems with electro-osmotic systems has been the depletion of electrodes acting as positive anodes embedded in negatively charged walls; platinized titanium and other resistant materials have largely overcome this particular problem.

1.5 REMOVAL OF SALTS

Any damp course insertion into a wall will, in time, create new patterns of moisture movement and, if effective, will almost inevitably attract accumulation of soluble salts. These may not only be below the damp course zone but above and even around old injection or feed points. Concentrations of salts may lead to local staining and spalling and can become quite unsightly. Consideration may need to be given to brushing down or even wetting and poulticing with absorbent clay packs from time to time. Volume 1, Chapter 6, 'Removal of salts ("desalination")' considers this matter further and describes suitable treatments.

1.6 COVERING DAMP WALLS

Much damage to wet masonry and especially to internal finishes has been caused by the mistaken application of dense impervious renders externally, and much money is wasted on waterproofing compounds internally. In the historic building context it is unlikely that the lining out or recovering of walls internally will be desirable, since the wall surfaces themselves are frequently of interest. If they are to be covered, traditional systems which may be effective include plasterboard linings on pressure-impregnated battens or patent corrugated bitumen lathing and plaster. Bitumen coatings tend to lose adhesion and a ten-year life is the maximum that could be anticipated. Replastering with a damp-resistant plaster may be effective. Decorations should be limewash or vapour-permeable emulsion and should in any case be delayed as long as possible to allow maximum drying of the plaster.

1.7 CONDENSATION

Terminology

Water will 'condense' from warm, moist air on contact with a cold surface. In sensitive conservation contexts it is necessary to measure accurately the climatic

conditions in which the condensation occurs. The following description is based on Technical Information Sheet 3 of the British Chemical Dampcourse Association (BCDA) (see References).

Air holds water in the form of water vapour (moisture). Warm air is able to hold more moisture than cold air. Air which contains its maximum moisture content is said to be *saturated.*

However, air does not always hold the maximum amount of air and is not always saturated. The *absolute humidity* is the weight of water vapour the air actually contains (g/m^3, in a given volume at a certain temperature).

The amount of moisture in the air is usually expressed as *relative humidity* (RH). This is the ratio of absolute humidity to saturation at the same temperature (for a given volume) that is, the ratio of actual moisture current to possible moisture content. The RH of air with a particular moisture content will vary with temperature, so as the air is cooled its relative humidity will increase.

	Air temperature °C	*Relative humidity %*
0.74% moisture content	20	50
	15	68
	10	95
	9	100

(BCDA, TIC3)

Below 9°C the air in the above example would be incapable of holding any more moisture and the surplus would be released as condensation. The temperature at which this occurs is known as the *dew point* and is dependent upon the amount of moisture in the air. The higher the moisture content the higher the dew point.

Condensation occurs mainly in cooler months when the external air temperature is low and external walls and windows are cold. Internal air which may, for instance, be warmed for the comfort of occupants, takes up moisture (for example, from paraffin and unventilated gas heaters, cooking and washing). This warm, moist air comes into contact with the cold surfaces and is cooled below its dew point. The phenomenon also occurs, of course, in unoccupied buildings, especially old buildings with massive masonry construction.

Dampness due to condensation is diagnosed from running water or droplets on cold surfaces, the presence of moulds, and an absence of hygroscopic salts. The wall temperature will be below the dew point of the air. Moisture readings on the wall surface will be uniformly high as distinct from a descending moisture gradient in the case of rising damp; the surface of the wall will have a high moisture level while the wall interior will usually be less damp. Moulds, usually but not exclusively black, will be found on the surface of paint or wallpaper, particularly in corners and behind large items of furniture where air circulation is restricted.

Overcoming condensation

Remedial action for condensation must involve both a lowering of moisture levels and the raising of cold surface temperatures. Heating the air alone is unlikely to be a satisfactory solution. A modest but constant background heat is preferable to intermittent heating, since this will help to maintain a higher ambient temperature in the fabric of the building. Improved heating and ventilation coupled with specific action in relation to cold spots will usually result in a significant improvement in conditions. Mould development within a building is unlikely to occur if the relative humidity is maintained below 70 per cent. Where heating and ventilation cannot be controlled sufficiently, a dehumidifier may be appropriate.

In old buildings with important surfaces and fittings a comprehensive survey of the condensation problem should be made including the measurement of air and wall temperature, relative humidity and the determination of the dew point of the air.

1.8 HYGROSCOPIC SALTS

Hygroscopic salts can draw moisture from the atmosphere, even when the relative humidity is less than 100 per cent. A wall that is contaminated with such salts can become damp, therefore, even under conditions where condensation would not otherwise occur. This often occurs in the evaporation zone of a wall which was subject to rising damp, and where a reservoir of salts has been deposited, but which has had its source of water cut off. The frequency of wetting/drying cycles and hence salt crystallization damage experienced by a wall containing hygroscopic salts can be high. Hygroscopic salts may also occur on chimney breasts due to the combustion of fossil fuel, or walls in coastal areas which are subjected to chloride mists and spray.

The presence of hygroscopic salts can be confirmed by analytical tests of wall materials. The most realistic solution to this problem is the removal of the salts by clay poultice. The transformation by chemical reaction of the soluble salts into insoluble salts (for example, the use of barium hydroxide to convert sodium sulphate to insoluble barium sulphate) may be considered theoretical rather than practical. Only in a museum context might it be practical to reduce the level of relative humidity to below the critical level for the salts involved, although useful controls can sometimes be effected by local heating at floor level.

1.9 DIAGNOSIS OF DAMPNESS PROBLEMS

The accurate diagnosis of the causes of dampness cannot be over-emphasized as an essential prerequisite for determining correct remedial action. In many cases the principal cause may be quite obvious, but in other cases several types of measurement may need to be made to achieve an accurate picture of the situation both within and surrounding a wall. The distribution of moisture and quantification of the degree of dampness is particularly important.

Methods of determining moisture

The measurement of moisture within masonry walls can be achieved by several methods with varying degrees of accuracy.

Carbide (chemical) method

A useful, accurate measurement of moisture is provided by the carbide method ('speedy' moisture tester). A standard weight of sample drilled from the wall is mixed with a set amount of calcium carbide power in a pressure vessel fitted with a gauge. The reaction between the powder and the water in the sample produces acetylene gas; the quantity, and hence pressure, produced is directly proportional to the moisture content of the sample. This method is easy to use on site, can be safely used with minimal instruction and gives results within 3–5 minutes per sample. It has a good degree of accuracy and requires less interpretation than any other method. Its accuracy is not affected by the presence of soluble salts. Comparison of results from drillings near the surface and in-depth will frequently distinguish between ground moisture and hygroscopic moisture or condensation.

Oven drying (gravimetric) method

This method involves transportation of drilled samples to a laboratory where the total moisture content is determined by oven drying. Moisture content due to the presence of hygroscopic salts can also be determined (see BRS TIL 29 'Diagnosis of Rising Damp'). While this method is highly accurate it requires the use of laboratory facilities and takes a considerable amount of time.

Electrical methods

Electrical methods of moisture measurement are of two types. Conductivity meters measure the electrical resistance of masonry, and the moisture it contains between two probes placed on the surface. Capacitance meters have an integral or separate sensor head which, when placed on the wall surface, measure the fringe capacitance in the sensor; this is influenced by the moisture content of the wall. Both electrical methods may not give a direct reading of moisture content. Confusion and incorrect diagnosis may arise where walls are of uncertain or mixed composition, where salts are present, where dampness is due to condensation or where the substrate being tested is itself a conducting medium (for example, contains a foil-backed paper). Electrical resistance meters can, however, give a good indication of zones of damp and assist in the early and quick identification of these.

Measurement in relation to condensation

Essential to the quantification of a condensation problem are measurement of air temperature, wall temperature, relative humidity and determination of dew point. This is a relatively simple process and involves one or more of the following instruments:

Relative humidity and temperature:

- Whirling hygrometer (psychrometer)
- Digital hygrometer/thermometer with air probe
- Hair hygrometer and a thermometer
- Thermohygrograph (long-term recording)
- Thermohygrometer (gives dew point reading as well)

Wall temperature:

- Thermometer with a surface probe.

Having determined the relative humidity and noted the air temperature, the dew point can be found by reference to a psychrometric chart or a similar table. If the wall surface temperature is at or below the dew point, condensation is possible and will be visible. In some situations it will be necessary to extend the period of monitoring over a period of several months to form an accurate picture of the environment and make a proper assessment of the risks involved to important surfaces.

REFERENCES

1 British Chemical Dampcourse Association (BCDA)
TIC 1 *The Use of Moisture Meters to Establish the Presence of Rising Damp*
TIC 2 *Plastering in Association with Damp-proof Coursing*
TIC 3 *Condensation*
TIC 4 *Methods of Analysis for Damp-proof Course Fluids*
TIC 5 *Chemical Damp-proof Course Insertion – the Attendant Problems*
TIC 6 *Safety in Damp Proofing*
TIC 7 *Chemical Damp-proof Courses in Walls – Detection Techniques and their Limitations*
Code of Practice for Installation of Chemical Damp-proof Courses (BCDA, 16a Whitchurch Road, Pangbourne, Reading, Berkshire RG8 7BP, Telephone: Pangbourne (07357) 3799).

2 British Standards Institution, *BS 743: 1970 Specification for Materials for Damp-proof Courses.*

3 Building Research Establishment, Digests:
Digest 77 *Damp Proof Courses*
Digest 110 *Condensation*
Digest 245 *Rising Damp in Walls: Diagnosis and Treatment.*

4 Building Research Establishment, Technical Information Leaflets:
TIL 29 *Diagnosis of Rising Damp*
TIL 35 *Electro Osmotic Damp-proofing*
TIL 36 *Chemical Damp-proof Courses for Walls*
TIL 47 *Rising Damp – Advice to Owners Considering Remedial Work.*

5 DOE Advisory Leaflets:
No 47 *Dampness in Buildings*
No 61 *Condensation.*

6 De Guchen, Gaël, *Climate in Museums*, ICCROM, Rome, 1984.

7 Duell, John and Larson, Fred, *Damp Proof Course Detailing*, London, The Architec-

tural Press, 1977. (This reference includes lists of Agrément Board certificates for most damp-proof course systems.)

8 Massari, Giovanni, *Humidity in Monuments*, ICCROM and Faculty of Architecture, University of Rome, 1977.

9 Torraca, Giorgio, *Porous Building Materials: Materials Science for Architectural Conservators*, ICCROM, 1982.

10 Richardson, Barry A, *Remedial Treatment of Buildings*, The Construction Press, 1981.

See also the Technical Bibliography, Volume 5.

2 THE ANALYSIS OF DAMP: CASE STUDY

2.1 INTRODUCTION

This chapter is a case study of the analysis of the levels and types of dampness present in the masonry of Burton Agnes Manor House, East Yorkshire, as carried out by the Research and Technical Advisory Service (RTAS) of English Heritage. The internal surfaces of the ground-floor masonry had been decaying for many years at a rate which gave increasing cause for concern. From initial inspections it was clear that dampness of various kinds was exacerbating this decay and that means of effecting some reduction in moisture and salts levels was required. In March 1984 a scheme was submitted to HBMCE for improving the situation by the introduction of a system of combined air and surface water drains around all wall and column footings. This proposal proved to be of great concern to the Inspectorate of Ancient Monuments (IAM) because of the extensive excavation and disturbance it would have produced in archaeologically sensitive zones. For this reason RTAS was asked to develop and consider more sensitive alternatives to the full air drain solution.

2.2 SURVEY WORK

In June-August 1984 survey work was carried out with the cooperation of the local area staff of English Heritage.

The facing materials of the internal ground floor of Burton Agnes Manor House are brick, limestone (clunch) and sandstone. Extensive repairs had been carried out over many years. Some of the oldest repair was patching with mortar; more recently many of the clunch voussoirs were replaced with concrete casts. Neither of these solutions was satisfactory and both had played some part in increasing the amount of deterioration. However, the decay was primarily associated with high dampness levels and heavy concentrations of soluble salts.

The RTAS/Area survey identified and quantified the dampness problem by:

- Determining the effectiveness of the disposal of rain and ground water. In the vicinity of areas of high masonry deterioration, ground water was found to be ponding on the outside of the building and a downpipe had been blocked in the past

- Moisture content readings were taken with a 'speedy' moisture tester in several parts of the building, so that a moisture profile for the building could be determined. Samples were taken at low and high points, on the surface and from the interior of the masonry so that the type as well as the amount of dampness could be identified. Temperature and humidity levels within the building were also recorded
- The extent of salt crystallization was plotted and a series of samples taken for analysis.

2.3 CONCLUSIONS

1 The moisture profile of the Burton Agnes masonry showed rising damp to be the principal type of damp. In almost every situation higher levels of moisture were found in samples taken at low levels than those taken at higher levels. Higher levels of moisture were found within the masonry as well as near the surface. Surface water ponding was considered the principal cause of this. It was considered that penetrating damp could not be ignored as a contributory cause especially at higher levels of the rear wall.

2 Weight gain analysis of the salt samples examined by BRE confirmed them to be hygroscopic. A possible source of salts was from the urine and excreta of animals which were once kept inside and outside the manor house. However, the major source of salts was found to be from Portland cement grouting which had been undertaken some fifteen years ago. Qualitative and quantitative analysis was confirmed by drilling.

2.4 REVISED PROGRAMME OF WORK

As a result of this analysis it was recommended that the programme of works for the manor house be revised to include:

- Adjustment of external ground falls to eliminate ponding
- Removal of as much salt as possible from the masonry

A phased programme of work incorporating these recommendations has commenced. External ground surfaces have been relevelled, to fall away from the building and towards a new system of sumps and drains. The surfaces have been paved with York stone slabs bedded on sand. The existing cobbles to the south of the building were retained and extended along the east side, the impervious tarmac having been cut back. The remaining sides of the building were flanked by a 900 mm margin of coarse rubble graded to fall away from the building. A 38 mm gap was left between all new ground surfaces and the walls. All ground

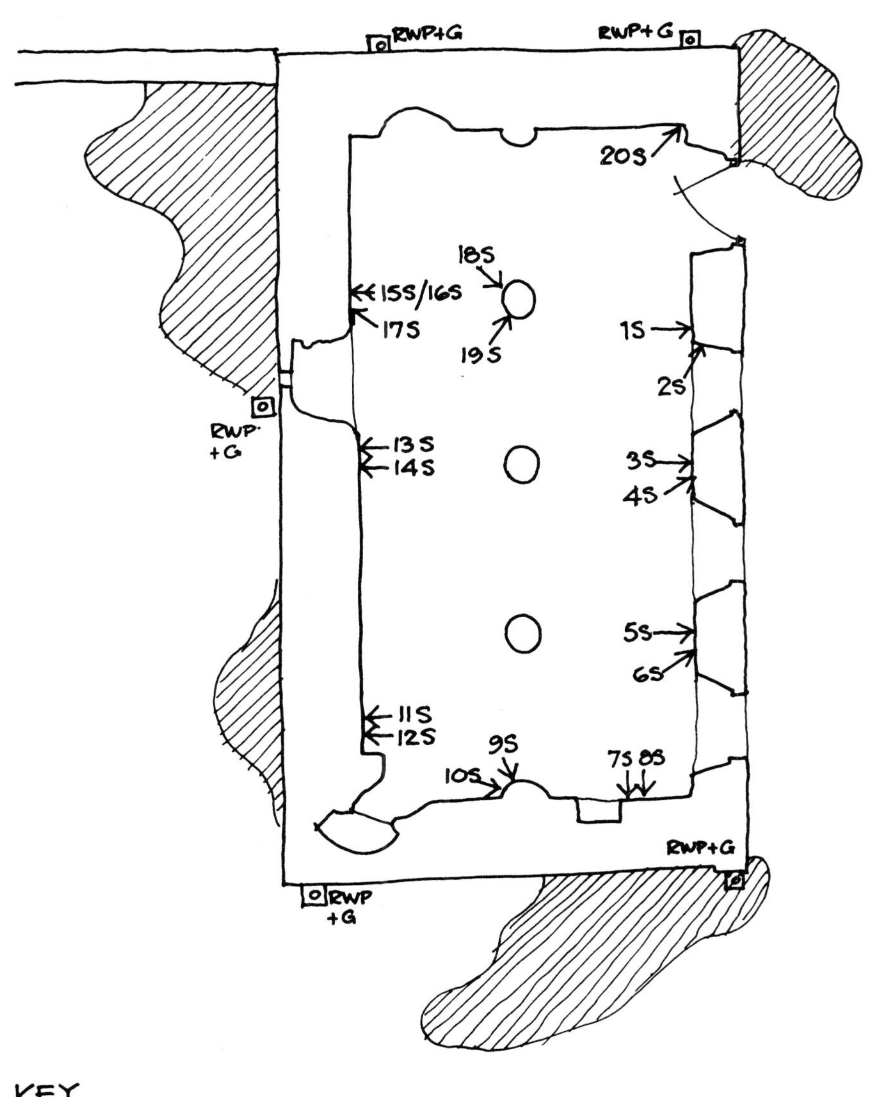

KEY

AREAS OF PONDING AROUND WALLS

3S → LOCATION OF MATERIAL SAMPLED FOR MOISTURE CONTENT MEASUREMENT

Figure 2.1 Surface water ponding and location of samples taken for moisture contents

Reference on plan	*Moisture content %*	*Position on wall*
1S 2S	5.2 4.6	low) high)
3S 4S	6.8 4.8	low) high)
5S 6S	16.0 8.0	low) high)
7S 8S	4.4 14.8	low) high)
9S 10S	11.2 8.4	low) high)
11S 12S	6.0 3.6	low) high)
13S 14S	14.8 6.4	low) high)
15S 16S 17S	7.4 (at 2″ depth) 12.6 (at 6″ depth) 10.0	low)) low)) high)
18S 19S	2.0 6.0	low) high)
20S 21S	8.2 6.4	low) high)

All paired readings with the exception of 7S and 8S (wall) and 18S and 19S (column) show a lower moisture content in samples taken from above the springing line than the samples taken from near the floor.

Figure 2.2 Moisture content readings taken with 'Speedy' moisture meter

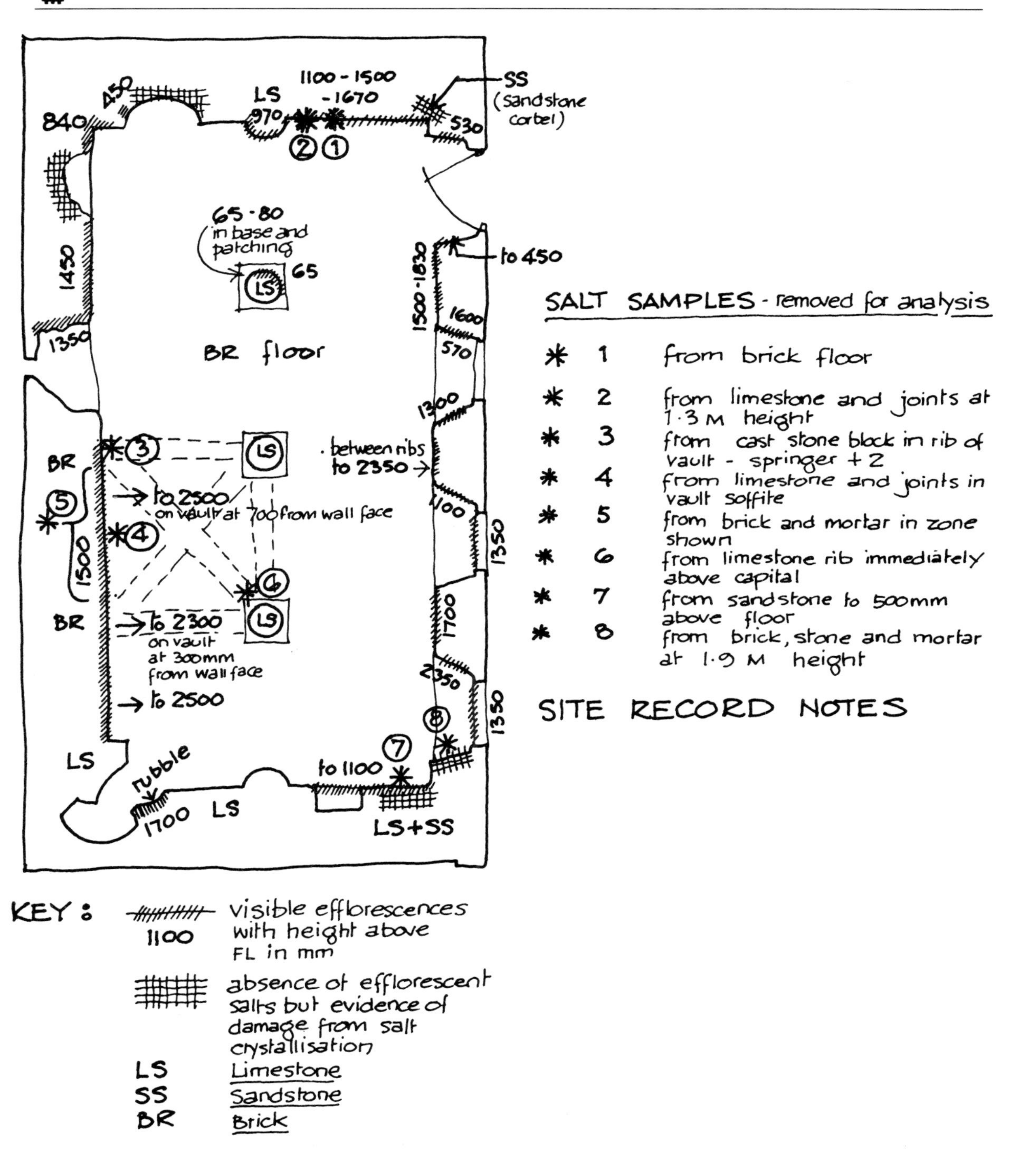

Figure 2.3 Extent of efflorescences and locations of samples for analysis

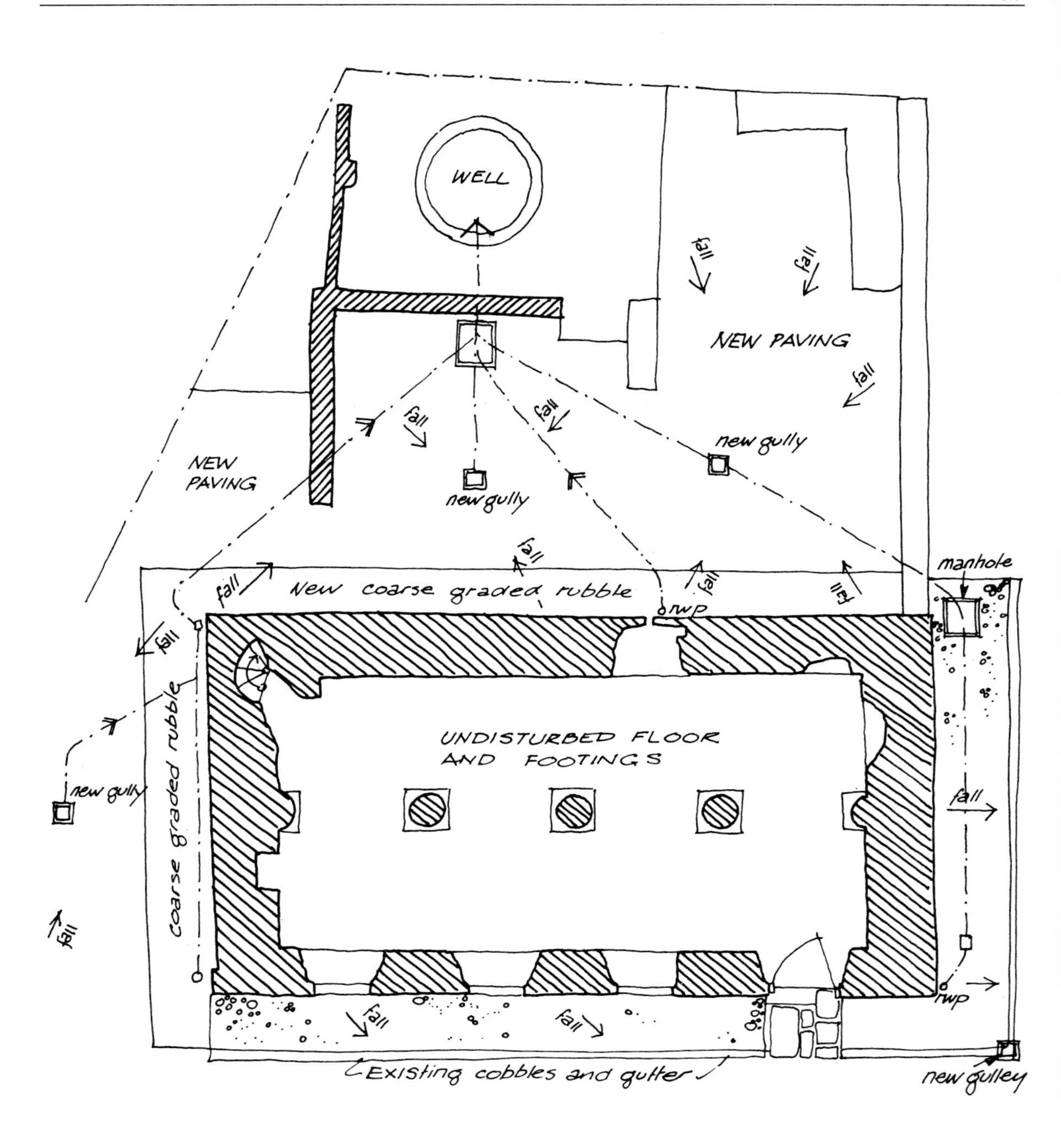

Figure 2.4 Revised proposals for rectification of damp

surfaces in immediate contact with the building now, therefore, discourage water retention and yet permit evaporation. These works are being accompanied by the poulticing using a sepiolite or attapulgite 50 mesh clay to remove as much salt as possible. Poulticing to the internal surface of the east wall has proved highly satisfactory. The method to be used is described in Volume 1, Chapter 6, 'The Removal of Salts from Masonry ("Desalination")'.

2.5 FUTURE WORK

The situation at Burton Agnes Manor House will require inspection and maintenance in the future. Annual poulticing will be necessary as more salts come to the surface, the size of the salts reservoir not being known. The success of the external regrading will need to be evaluated and the desirability of a combined air and land drain to the north and west walls determined.

3 MORTAR ANALYSIS

3.1 LIMITATIONS OF MORTAR ANALYSIS

The study of existing mortars in historic structures is an important aspect of building conservation. This chapter describes the potential and limitations of mortar analysis. The choice of a mortar must firstly relate to the type and condition of the masonry and secondly to the degree of exposure. This choice is primarily based on a knowledge of the properties of various mortars and is not arrived at by analysis of the existing mortar alone. There may be good, archaeological reasons for wanting to establish the identity and proportions of constituents in an old mortar and simple separation of aggregates may be useful in identifying likely sources of aggregates for matching purposes. There are, however, limitations which should be understood and mortar specifications should not be based on the simple breakdown analysis of a sample. Analysis requires interpretation and there are important factors which affect the condition and performance of the mortar being sampled which analysis will not reveal. Examples of such factors are the original water:binder ratio, the rate of drying out, the method of mixing and placing and the cleanliness and conditions of the aggregates.

There are practical difficulties, too, in isolating and identifying constituents. For instance, calcareous aggregates will be digested with the calcareous binder material in acid and present a misleading binder:aggregate proportion (the occurrence of old mortar crushed down and reused as aggregate is a notorious problem of this kind); or clay minerals, present as impurities, may not be readily distinguishable from the silicates present in an hydraulic cement. An additional difficulty is accurate matching of an old clamp-fired lime, well mixed with fuel and kiln slag, with a modern lime produced in closely controlled conditions and delivered as a very pure hydrate.

The method of mortar analysis selected depends on the information required. Sufficient data may be provided by in situ visual analysis or simple on-site visual, physical and/or chemical testing. Laboratory analysis can provide additional information which may not be necessary for the task at hand.

The analysis of mortars is, however, a specialized field. Even for the simplest method, experience is required for the identification of materials and the correct interpretation of evidence.

Mortar analysis and dating of structures

The analysis of old mortar is the analysis of changing technology. Only rarely can it be used to provide a specific date of construction. Laboratory analysis and expert interpretation will provide the most detailed and accurate results, but by themselves they are unlikely to give much indication of the dates when a mortar was prepared. The dating of masonry walls should not be attempted or expected from mortar analysis alone. A combination of evidence is necessary. A thorough examination of documentary sources relating to a building/structure and its site should be carried out by an experienced person with training in historical research. This should be correlated with analysis of the fabric, one aspect of which is mortar analysis. Further examination of the fabric should include interpretation of the method of construction and the manufacture of its bricks and/or preparation of its stone.

3.2 METHODS OF ANALYSIS

On-site analysis

On-site visual analysis by an experienced person will provide a good indication of the general components of a mortar, particularly where a hand-held X10 magnifying glass is used. The binder, aggregate and other large particle inclusions can usually be identified. Gentle scraping of a weathered surface may be necessary to reveal the unweathered mortar. This inspection will also provide useful information about the construction of the masonry, such as the original joint profile, the condition of the original mortar and whether the joints have been repointed. It is usually advisable to remove a representative unweathered sample of mortar and inspect it with a X10 magnifying glass in good lighting conditions. This sample should be inspected further after it has been disaggregated (crushed, but not ground).

At the completion of these inspections it should be possible to know whether the binder is predominantly lime-based or cement-based (Roman or Portland) or whether it contains a substantial proportion of clay or loam. The type and general characteristics of the aggregate as, for example, rounded or angular sand grains, crushed stone and brick dust should be revealed. Larger inclusions such as gravel, unburnt shell lime, shell aggregate and kiln slag will also be identified. The ease with which the mortar is scored with a fingernail and knife and the removed sample broken will help to identify the presence of any hydraulic constituent.

Simple chemical analysis

Chemical analysis is usually required to determine the proportions of mortar constituents. Several professional mortar analysis services are available in the UK.

RESEARCH, TECHNICAL AND ADVISORY SERVICES

ENGLISH HERITAGE

MORTAR ANALYSIS SHEET

SITE/MONUMENT:
Kenwood House
LOCATION OF SAMPLE:
Service Wing-mortar

SAMPLE TAKEN BY:
?
VISUAL DESCRIPTION:
A beige-grey mortar, crushed chalk up to 10 mm, dark quartz aggregate, prominent, large pieces of brick (from wall?). Portland cement repointing attached to some pieces. Black specks in beige mortar - general appearance.

RTAS REF: 692/34

SUBMITTED BY:
M. J. Stock, London Division
PURPOSE OF REQUEST:
Replacement mortar

ANALYSIS:

DISAGGREGATION:
Not all pieces of P.C. could be removed. Piece of brick crushed inadvertently.
DISSOLUTION:

FILTRATION:

DRYING:
The surface zone of the mortar did not dissolve.

WEIGHT (DRY): 17.37g

WEIGHT: 10.06g

AGGREGATE DESCRIPTION:

SIEVE ANALYSIS:

5.0mm	
2.36mm	**5%**
1.18mm	**5%**
600microns	**10%**
300microns	**30%**
150microns	**40%**
<150microns	**10%**

CALCULATIONS:
$\frac{10.00}{17.37} \times 100 = 58\%$ aggregate

Original mix (approx.)

1 lime : 1 aggregate
to
3 lime : 4 aggregate

BY: Nicola Ashurst **DATE:** 3rd November, 1987

Figure 3.1 (a) Mortar analysis sheet

RESEARCH, TECHNICAL AND ADVISORY SERVICES

ENGLISH HERITAGE

==

MORTAR ANALYSIS SHEET

==

SITE/MONUMENT:
Kenwood House - Service Wing

RTAS REF: 692/34

LOCATION OF SAMPLE:
Mortar

SUBMITTED BY:
M. J. Stock, London Division

SAMPLE TAKEN BY:
?

==

RESULTS AND RECOMMENDATIONS

==

The sample submitted was a beige mortar with crushed chalk pieces (up to 10 mm) and dark quartz aggregate prominent. The mortar appeared to be a traditional well mixed lime aggregate mortar. There were several pieces of Portland Cement - based repointing material attached and these were not included in the analysis.

The analysis revealed the original mix to be approximately 1 part lime (putty) to 1 1/4 parts aggregate. The surface of the mortar was heavily sulphated. The aggregate, a fine quartz sand was graded in the range 2.36 mm to less than 150 microns with most emphasis on 300 and 150 microns.

There were also some pieces of coal and charcoal (probably from the lime burning in the kiln).

In spite of the high proportion of lime determined by analysis we would not recommend that this should form the basis of a replacement mix, because of the purity of modern lime. To match the overall appearance and strength properties we would suggest a 1 : 1 1/2 lime putty : aggregate mix using the attached sand sample for colour matching (Please note the sample contains whole and fragmented pieces of the sulphated skin, which should be disregarded for the purpose of aggregate matching).

We suggest that this lime sand stuff is gauged with HTI powder in the proportions 1 part HTI to 6 parts coarse stuff within one hour of use.

BY: Nicola Ashurst
Research Architect

DATE: 3rd November, 1987

ENCLOSURES:
Aggregate for matching

Figure 3.1 (b) Mortar analysis sheet – results and recommendations

The basic principle of analysis is first to dissolve the lime binder in acid, then to separate the aggregate (sand, brick dust, crushed stone) and the fines (cements, fine brick dust and crushed stone), and thereby determine their proportions. Only simple laboratory facilities are required to undertake the procedure which is as follows:

Examination and dissolution of the binder

1 Collect an unweathered sample of about 40–50 grams. Examine it and record characteristics such as colour, texture, aggregates, inclusions and hardness (scratch resistance). Retain half as a record

2 Powder half the sample with a mortar and pestle. Dry at 110°C for 24 hours and then weigh it with a balance (to an accuracy of 0.1 g)

3 Place the sample in a glass beaker and moisten it with deionized water. Then immerse the moistened sample in a 10–15 solution of hydrochloric acid to dissolve the lime binder. The mixture will effervesce as CO_2 is given off. Carefully observe the reaction through the side of the beaker. (Safety glasses should be worn). The mixture should then be stirred with a glass rod to make sure the reaction is complete

4 Weigh a piece of filter paper, place it in a funnel positioned over a large flask

5 Add a few drops of hydrochloric acid to the sample to ensure complete acid digestion of the binder and stir. Add water to it slowly and swirl with a glass rod to suspend the fines

6 Pour the liquid with the suspended material through the filter, being careful to keep the solid particles at the bottom of the beaker. Add more water and repeat the swirling and pouring until the water added to the beaker remains clear

7 Dry the fines collected on the filter paper and weigh. Determine the weight of the fines

8 Wash the sand with water several times and leave to dry for 24 hours. Weigh the dry sand.

Alternatively: Steps 6–8 may be combined. All the sand and fines from the beaker may be poured into the filter and dried and weighed together (see step 11)

9 Express the amounts of sand and fines as a percentage of the whole sample. The amount of dissolved binder is calculated by subtracting the sand and fines weights from the weight of the original sample. The weights determined will give the proportions of binder, fines and aggregates of the original mix. Allowances must be made for the loss of any calcareous aggregates dissolved with the binder. The results of the analysis can be recorded on a sheet such as the one at Figure 3.1

10 Inspect the colour of the dried fines. Simple inspection of this kind is

normally sufficient to identify clay (yellow, plastic when wetted), brick (red/brown), cement (grey), sand (almost any colour, gritty to touch)

11 More accurate examination must be made with a binocular microscope to determine colour, particle shape and material types. Sieve through standard sieves to determine particle size distribution expressing the amount of each particle size as a percentage of the whole. (*Note*: the acid may have changed the colours of the sand)

X-ray diffraction

Sometimes more sophisticated techniques are needed, mainly for historic mortars research, to provide more specific information than separation and sieving of constituents. In this case a sample is submitted to a laboratory where a portion of it is ground to a homogeneous powder and a mineral analysis conducted by X-ray diffraction (XRD). The sample is irradiated and the crystal planes of the materials in it reflect the rays.

Lime, sand, Roman cement, Portland cement and pozzolanic additives such as trass can be identified by this method of analysis, as each has a different crystal diffraction pattern. XRD provides evidence of clay in a mortar and is able to identify the type of clay present. The experience of the operative in interpreting such results is very important.

3.3 SAMPLING PROCEDURE

Particularly where mortar analysis is part of a programme of archaeological and historical research, a thorough sampling procedure is required. The objectives of the sampling should be defined well in advance. The number and size of the samples should be the minimum necessary to gain the required information without doing damage to the historic structure. The sampling procedure should include the following guidelines:

1 Sampling should be done by persons well acquainted with a building or its remains, to ensure that a proper programme of sampling is prepared and the samples receive proper interpretation. It is important to involve the analyst in the sampling operations

2 The sample should preferably be in the form of lumps, not crumbled or powdered. The quantity usually required for comparative analysis and for reference material is about 40–50 g (2 oz), preferably in one or a few compact fragments (half for analysis, half for reference)

3 The exact position (not just the location) from where the sample was taken must be accurately recorded

4 To make certain that a particular kind of mortar is typical for a certain wall, at least three samples should be taken from different parts of that wall

and analysed separately. If they prove to be identical within limits of practical deviations, their composition and properties can be considered as typical

5 The sample must be clearly and thoroughly labelled

Recording

Figure 3.1 illustrates a typical mortar analysis sheet on which the analysis of a lime render sample has been recorded. The sheet shows the kind of information which would be received from a laboratory after examining a mortar by chemical analysis and grading of aggregates. In this case the most useful contribution made by the analysis to work on site was the identification of aggregates which enabled them to be matched with some accuracy. Five per cent HTI powder was added to the repair mix as the wall surface, originally internal, was to remain exposed.

REFERENCES

1 Ashurst, John, *Mortar Plasters and Renders in Conservation*, Ecclesiastical Architects and Surveyors' Association, 1983.
2 Malnic, Nicola, *Masonry Mortars in Historic Buildings*, unpublished dissertation, Master of the Built Environment (Building Conservation), University of New South Wales, Sydney, 1983.
3 Teutonico, Jeanne-Marie, *Architectural Conservation Course, Laboratory Specifications, Tests and Exercises*, ICCROM, Rome, 1984.

See also the Technical Bibliography, Volume 5.

4 POINTING STONE AND BRICK

4.1 INTRODUCTION

The deterioration of mortars is a common concern of all who find themselves responsible for the maintenance of an old building of whatever form of construction. Joints weather out, cracks and loss of adhesion result from building movement, and water penetration and decay of adjacent bricks and stones take place. There are, however, two ways in which deterioration processes are advanced; one is neglect, the other is the wrong kind of remedial treatment. Thus it is vital that proper consideration is given to every situation and that snap judgements are not made based on superficial appearances.

Pointing is the process of filling the outer part of the joints in stone and brick masonry where the bedding mortar has been deliberately left or raked back from the surface or where the original mortar has weathered back from the face.

Re-pointing is the re-filling of the outer part of the joints where there was an original pointing which has weathered out. This chapter deals with the considerations and procedures involved.

4.2 CONSIDERATIONS PRIOR TO COMMENCEMENT OF POINTING

Examination of the masonry

All pointing and re-pointing on historic masonry buildings should be preceded by a detailed visual analysis of the wall to determine:

- The nature and properties of the masonry units, for example, weak porous bricks, hard extruded bricks, sandstone or granite
- The characteristics and constituents of the original mortar and its relationship to the masonry units
- The mode of construction of the wall, that is, the bonding, width of joints, joint profile and texture of the pointing.

This illustration is a striking example of the dramatic visual change re-pointing can bring about. Whilst this, in itself, is acceptable when necessary and when properly executed, it is very much to be regretted when the original mortar is in good condition. This eighteenth-century brickwork in English bond is in almost perfect condition (right hand side). The mortar is weathered, but sound. On the left of the picture unnecessary, weatherstruck pointing has spoilt the wall. The first question must always be – does the wall need pointing?

Matching mortar properties

On some buildings a simple visual analysis of the mortar by an experienced conservation practitioner may be sufficient to determine an appropriate match for the new mortar. The exact physical and chemical properties of the original mortar are not always of major significance as long as the new mortar:

- Matches the original mortar in colour, texture and detailing
- Is softer, in terms of compressive strength, or more porous than the brick or stone
- Is as soft or softer, in terms of compressive strength, as porous or more porous, than the original mortar.

However, in many cases the historic significance of a building will require a mortar analysis to be comprised of an on-site visual analysis and a laboratory examination of the mortar. Laboratory analyses may provide binder:aggregate ratios and sieve grading of aggregates which can be useful indications of contemporary practice and sources of constituents. They may also assist the reconstitution of a matching mortar. (See Chapter 3, 'Mortar analysis'.)

When planning and executing refilling of joints it is always critical to remember that the mortar must be so constituted and so placed that it will act as an integral part of the masonry wall. It is also critical that the pointing mortar is physically compatible with other mortar and masonry materials and that the mortar and joint treatment are historically accurate and aesthetically appropriate to the appearance of the structure.

Matching mortar colour and aggregates

The colour of the aggregates, especially sands, is important because it is this, primarily, which gives the mortar its overall colour. The grading of the sand particles is also significant because of its effect on the properties of both plastic and hardened mortar and its contribution to the texture of the repointed mortar surface.

The aesthetic tradition of stonework is generally different from that with brick, as the joints are usually close in colour and texture to the stone. Every attempt should be made to follow this tradition working on the basis of surviving evidence. With brickwork, the final colour and texture of the mortar is particularly important as it forms up to 30 per cent of the surface area of the wall, so in time, the coloured texture of the sand will dominate.

The use of pigments should not be relied upon to achieve a satisfactory colour match. Although it is acknowledged that there may be cases where it is very difficult to obtain a satisfactory match without these, every effort should be made to locate sands with a satisfactory staining property as part of the total aggregate.

A new pointing mortar should match the colour of unweathered interior portions of the original mortar of a wall. Samples should be made, allowed to cure and their broken surface compared with that of a sample of the original mortar.

4.3 MORTAR JOINTS

Preparing the joint

If the surface of mortar joints has weathered out to the extent that the face of the stones of bricks is vulnerable to damage, or that water can lodge and penetrate, or that support is inadequate, then a matching mortar must be introduced. In general, the original mortar joints of historic work will not have been bedded and pointed in separate operations, with notable decorative exceptions, but filled full and struck off flush as the work was raised. If the stones or bricks have retained their sharp arrises then the joints should be filled flush again, unless there is specific evidence that the joint face was profiled in some other way. Long years of weathering, however, will normally have blunted these arrises and sometimes all the original face of some stones will have spalled off (see Figure 4.1). Flush filling in such a situation will greatly increase the apparent width of the joint and, therefore, great care must be taken to keep the face of the new mortar within the original joint width, however far back that may be.

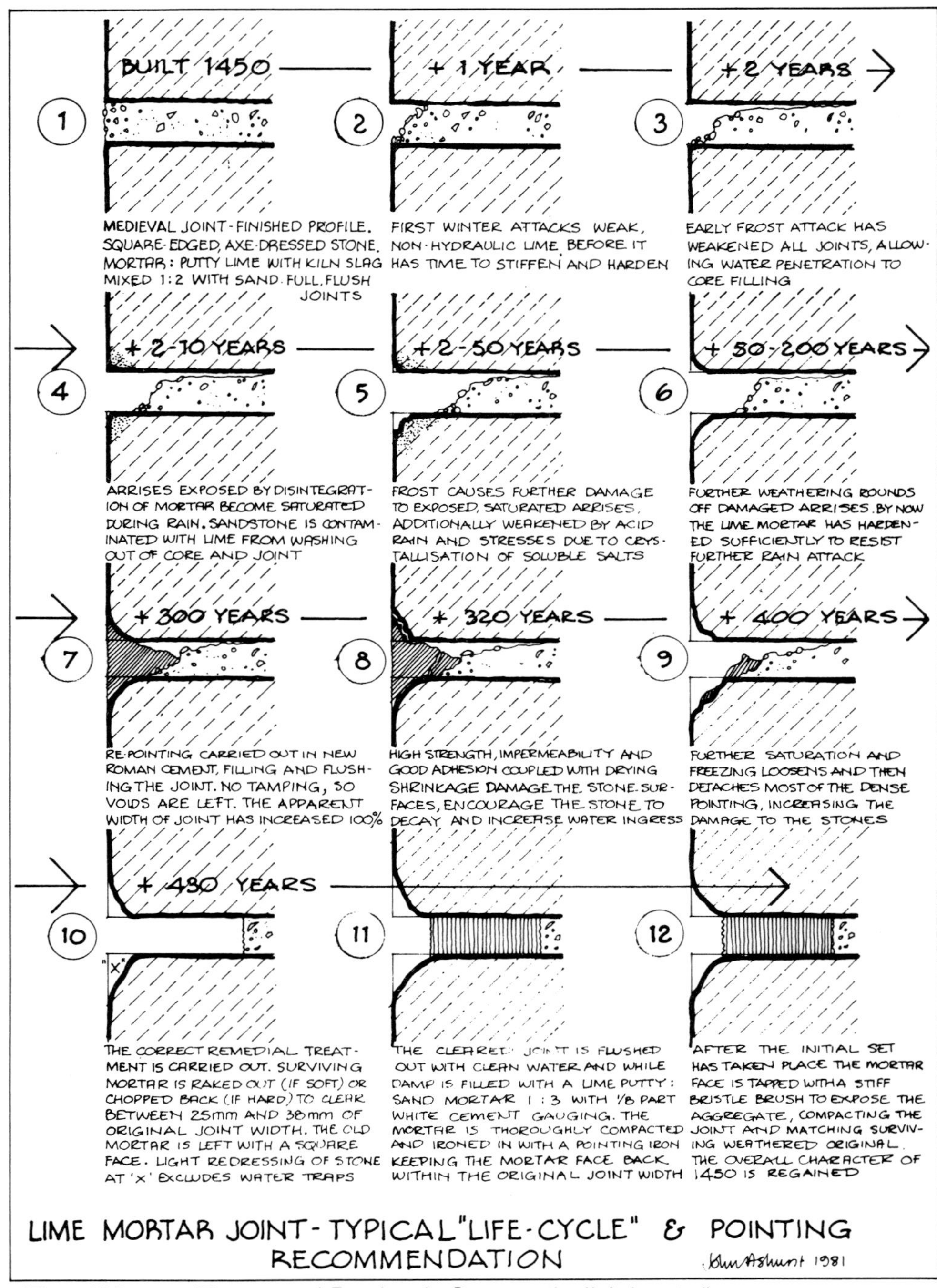

(From "Mortars, Plasters and Renders in Conservation" Ashurst, J)

Figure 4.1 Lime mortar joint: typical life-cycle and pointing recommendation

The correct procedure of cleaning out and re-filling is well known but unfortunately not widely practised. As a general rule joints should be cleaned out to a minimum depth of 25 mm (1 in) and never to a depth less than their width; but wide joints, especially those liable to exposure to extreme weathering, should be cut out to a minimum 38 mm ($1\frac{1}{2}$ in) or even 50 mm (2 in). Sometimes the mortar has disintegrated to such an extent that the joints are largely empty, in which case they must first be deep tamped and, if necessary, hand grouted to fill the joint to the required depth for pointing. If tamped or grouted mortar comes closer to the face than 25 mm–38 mm it must be cut back to the proper depth and to a square face before pointing.

Raking out may be a simple operation without risk to the fabric where mortar is substantially decayed, but it is over-simplifying the situation to say that the joint does not need re-pointing if it requires cutting out. Not infrequently the face of a lime joint has been lost early in its lifetime, before sufficient drying out and carbonation had taken place to enable it to resist the winter, but the more protected mortar has survived to become extremely hard. The empty joint at the face may be too much of a risk to leave alone, and additional cutting out may be necessary to achieve enough depth for pointing. More often, cutting out (as distinct from raking out with a knife blade or bent spike) is necessary to remove dense re-pointing of an earlier period, especially where this mortar – fortunately usually shallow in depth – is causing problems because of its high strength, impermeability and tendency to trap water behind it and accelerate the decay of stones and bricks.

Cutting out should be achieved using quirks, plugging chisels, long necked jointing chisels and toothed masonry chisels, with a $2\frac{1}{2}$ lb club hammer but never with chisels which tend to wedge the joints and cause spalling (See Figure 4.2). Impact should be at an oblique angle to the joint face, not directly onto it. Hacksaw blades and masonry saws may also be used and drilling with masonry drills is a useful way of creating an initial breach into a strong mortar. Exceptionally, small carborundum or, better, diamond discs may be used in cutting out, but usually only on regularly coursed work with level beds where running rules can be fixed to the wall as guides for the power tool. The risks of over-running are obvious, and extreme caution must be used not to cut into the masonry or increase the width of the joint. Because of the high risk of damage when discs and power tools are used, it is usually advisable to ban their use on historic brick or stone.

All cutting out should leave a clean, square face at the back of the joint to provide optimum contact with the new mortar. Time for cutting out, which may be considerable, must be properly programmed. Inadequate cutting out and the inevitable shallow depth of pointing which follows is a complete waste of time and money. In this context a warning should be given against the use of proprietary 'cut and fill' methods. There are no cheap, fast alternatives to craft work.

Cleaning the joint

The prepared face should be carefully cleaned out with a soft or stiff bristle brush and thoroughly flushed out with clean water, avoiding unnecessary saturation.

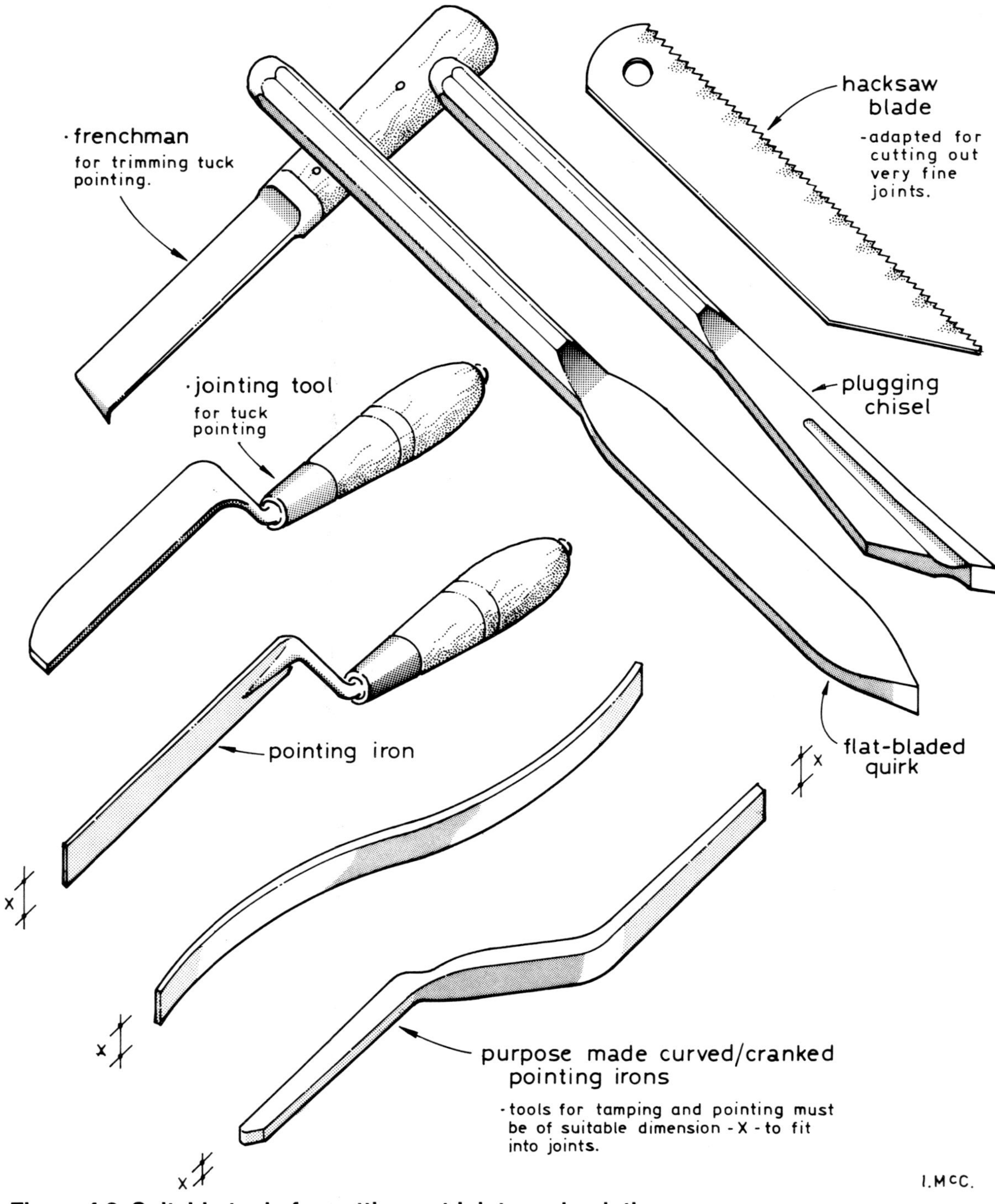

Figure 4.2 Suitable tools for cutting out joints and pointing

All dust and loose material must be removed, working from top to bottom of the wall. If old, weathered-out joints have been colonized with algae or lichens a biocide must be used on the dry surface as part of cleaning out.

Other preparation

In some situations, especially where fine joints are to be filled, it may be advisable to protect the faces of bricks or stones with masking tape or masking paint (latex). These should not be left on the surface for longer than necessary, especially when exposed to hot sun.

Filling the joint

(See also page 37, 'Filling fine joints').

If the joints have dried out after cleaning they must be re-wetted before placing the new mortar, to avoid undue suction taking too much water too soon from the mortar. The mortar is pushed into the joint from a board and ironed in with the maximum possible pressure. Pointing trowels are in common use, but it is regrettably unusual to see pointing keys, which can be improvized to suit the particular work in hand. These may be cranked bronze or steel flat, or beaten out rod or even wood (See Figure 4.2). Even if the mortar is placed in the joints using a gun, which is increasingly popular where large areas are involved, the mortar must subsequently be packed with a pointing key. The function of these keys is to push the mortar evenly into the joint for the full joint width; this they can do because they fit into the joint and do not try to achieve compaction from the surface alone. In irregular work this is particularly important.

The mortar face should be filled flush, or slightly recessed to avoid spreading the mortar over the face of the masonry, or struck and lined out as required. Finished work should be protected from direct sun and rain until the face has dried and hardened. If a weathered appearance is desired to match existing surviving work, a roughened texture can be produced after the initial set of the mortar has taken place by light spraying, by stiff bristle stippling or by dabbing with coarse sacking. Experience, but above all an understanding of what is required on the part of the mason or bricklayer is essential. Of the above techniques, stippling with the ends of the stiff bristles in the brush is probably the most universally successful. The bristles should not be dragged across the face but tapped against it. Timing is critical, and no specification can substitute for experience. If this technique is applied too soon mortar will be removed too easily and the bond forming between mortar and stone or brick will be disrupted; if too late, it will be difficult to make an impression and too vigorous effects may be made with wire brushes or masons' drags to achieve the effect. As a guide, the surface setting time of a range of mortars on different sites is shown in Table 4.1.

The times shown in the right-hand column indicate when surface treatment was applied to the joint face. Apart from leaving a pleasant, weathered appearance the rough textured joint tends to assist the wall to dry out and to concentrate wetting and drying activity (provided the right mortar is used) in the joints.

Table 4.1
Setting times for a range of mortars

Site		*Mortar*			*Surface set*
		Cement	Lime	Sand	Hours
Ewenny Priory	hard limestone	–	1	3	36
Caernarvon	limestone	–	1	2	24
Kidwelly	hard limestone	$\frac{1}{8}$	1	$2\frac{1}{2}$	18
Old Wardour	glauconitic limestone	$\frac{1}{8}$	$1\frac{1}{2}$	4	14
Okehampton	sandstone	1	–	5	12
Castell Coch	hard limestone	1	1	5	4

There is a danger with this type of weathered finish that the joints may take on too distinctive a character. The old style of washed grit pointing which used to be carried out by the Ministry of Public Building and Works on ancient monuments was often a work of art but tended to become an end in itself. Over-pretty work, with mortar kept back to emphasize the outline of every stone can look very self-conscious and rather odd in an area where the tradition is to slurry over joints in rough rubble masonry to produce a flush face. A study should be made of surviving masonry to avoid the worst mistakes of this kind.

More serious and more common is the error at the opposite end of the scale. Thick strap pointing raised proud of the wall will positively shorten the life of the masonry, especially when the surface is already weakened with decay and will totally alter the appearance of the masonry. Strap pointing is not suitable for any historical masonry.

Cleaning off

Keeping the work clean is part of the skill of the mason or bricklayer but occasional staining from mortar is an inevitable hazard. Sometimes washing and brushing down is sufficient to remove recent material, but if the traditional 10 per cent concentration of hydrochloric acid is used (or a proprietary product based on this acid and a surfactant) the masonry surfaces must be pre-wetted to limit absorption and the acid must be thoroughly washed off afterwards. The biggest problem is keeping gauged brickwork clean, and every effort must be made to protect the vulnerable surface of these bricks from mortar spread and droppings when filling the joints.

The importance of craftsmanship and supervision

It is important that the mortar mix and the joint preparation procedures of a pointing job are specified clearly. This will enable fair tendering and will give the contractor a clear picture of what is required of him.

It is of even greater importance that the workmanship of the cutting out, preparation and placing of the mortar is carried out well. The following aspects are critical to this:

- The cutting out must be deep enough and the joints cleaned and dampened to receive the new mortar
- The mortar must be consistently of the specified constituents, well rammed and beaten, contain a minimum of water and any hydraulic additives must not be added more than an hour before the mortar is placed
- The mortar must be placed firmly against the back of a joint with the appropriate tools, tamped firmly, and receive the agreed surface treatment
- The quality of workmanship throughout the job must be consistently good.

It is therefore recommended that supervision of a pointing job includes the following:

- Approval of all cutting and clearing out before pointing
- Approval of the lime putty, aggregates, gauging material (if any) and approval of a dried mortar sample which should remain for reference during the job
- Approval of the coarse stuff prior to gauging and approval of the method of gauging
- Approval of all tools used
- Approval of a test area of pointing and the supervision of the selection and undertaking of this. The test area should be retained throughout the job for comparison
- Approval of the skill levels of all workmen permitted to undertake the work
- Cutting out of new pointing at positions selected by the supervisor.

Special joint treatment

Common historic variations on the flush joint were the beak or double struck joint, and later, in the eighteenth century especially, 'joints jointed' and tuck pointing. Where evidence of these survives it should be the pattern for the remedial work, unless the face of the masonry has decayed to the extent of making a finished joint of this kind appear nonsensical. Although there are historical precedents for overhand struck joints, the modern weathered struck joint and the 'bucket-handle' joint profiles are normally inaccurate and visually disturbing in an old wall.

Filling fine joints

Unfortunately there are some joints which are so fine that it is quite impracticable to talk of pointing them. Stone masonry with fine joints may be worked to true, flat beds but more commonly will be hollow bedded to allow for a generous mortar fill and a fine joint at the face. This was an economic way to produce a fine quality appearance. Injection grouting of these joints is notoriously difficult, as the grout tends to disappear without any effect into the wide, open joints behind the face.

Preparation

There are three methods for filling such joints, all of which must be preceded by careful cutting out with a fine-toothed saw blade or hooked knife blade. The inserted blades will indicate when an adequate depth has been cut out. It is advisable to cut out to a depth of 25 mm. The joints must then be flushed out using a large (30 cc) hypodermic syringe and clear water until the water runs out clean. A backing 'rod' may be necessary to contain the new mortar when it is introduced by a grouting technique and may be desirable when the mortar is introduced as a putty. The simplest form is a length of fine waxed string, or string coated in petroleum jelly. Stretch the string and twist it several times before pushing it into the back of the joint with a knife blade. As the string tries to 'unwind' it will close off the potential escape of the mortar.

Method 1: putty sandwich (See Figure 4.3.)

Cut two sheets of thin stiff plastic film approximately 150 mm (6 in) × 125 mm (9 in) for the bed joints and two sheets approximately 150 mm (6 in) × 75 mm (3 in) for the vertical joints. Prepare a finely screened lime putty by passing it under the edge of a trowel on a glass sheet after sieving (ordinary lime or hydraulic lime putty). Spread the putty with a trowel on the plastic sheet and lay the second sheet on top to form a sandwich. The putty should form a 50 mm (2 in) wide band between the sheets. This sandwich is then manoeuvred into the back of the clean, damp, cut out joint. Open the two leaves of plastic where they project from the joint and lay the blade of a flat ironing tool between them against the wall face. Keeping firm pressure on the blade, withdraw first one leaf of plastic and then the other. The putty will remain in the joint. Trim off any surplus or spread of mortar with a clean pointing trowel.

Method 2: mortar injection

After the preparation described above and using a small pointing trowel or spatula, seal the outer face of the joint with 3–5 mm of mortar, squeezed in from the face. Using a 10 cc hypodermic syringe, inject a grout consisting of 4 parts hydraulic lime, 1 part finely powdered brick dust and twelve parts gauging liquid. The gauging liquid consists of water to which one half part of acrylic emulsion has been added. Push the hypodermic needle through the thin mortar seal and inject slowly. At approximately 150 mm (6 in) centre holes should be opened to check the grout flow and allow air to escape. These holes can be closed up with a small squeeze of lime putty. The following day the lime can be cleaned off the surface with a hand spray and small bristle brush. This method is not recommended for brickwork and especially not for gauged brickwork because of the risk of lime staining, but is suitable for fine jointed stone ashlar of almost any kind.

ADHESIVE TAPE METHOD

1. *Decayed mortar raked out with hacksaw blade. Care must be taken not to damage arrises. Joints flushed out with clean water from hypodermic syringe.*
2. *50 mm heavy duty adhesive tape (carpet type) applied over joint. Tape slit with sharp knife along centre of joint.*
3. *Mortar introduced into damp joints and compressed with pointing iron through slit in tape.*
4. *Adhesive tape carefully peeled away.*

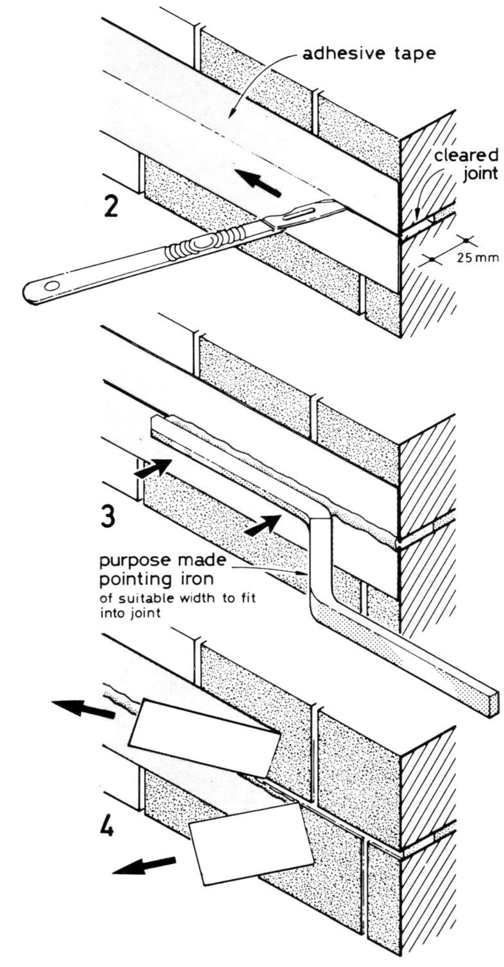

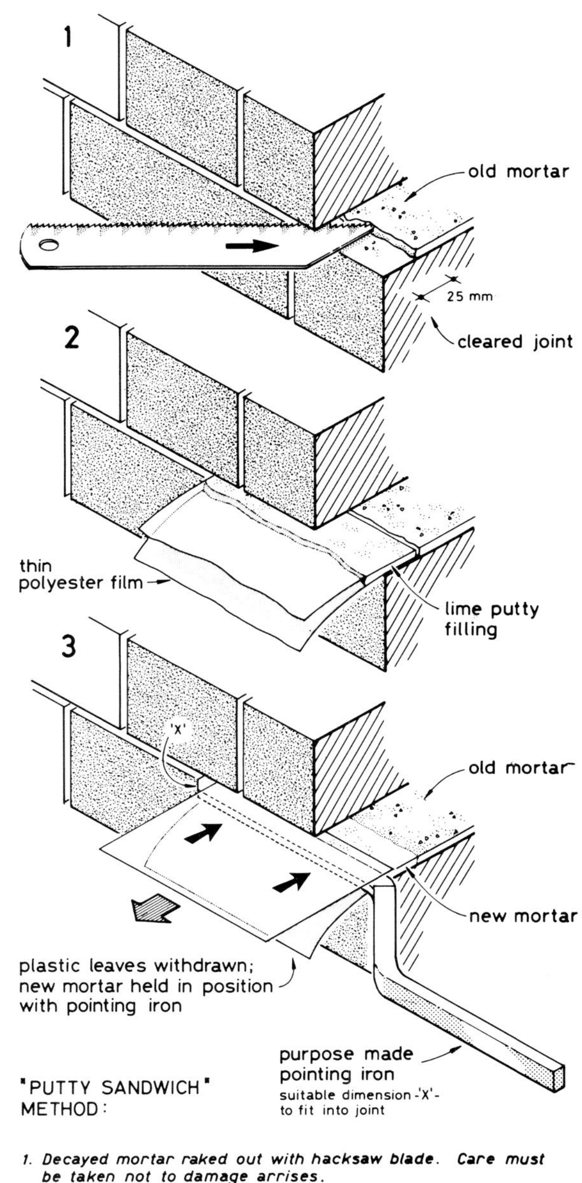

"PUTTY SANDWICH" METHOD:

1. *Decayed mortar raked out with hacksaw blade. Care must be taken not to damage arrises.*

   Joints flushed out with clean water from hypodermic syringe.
2. *Finely screened lime putty spread in band on sheet of thin polyester film. Second sheet laid on top to form sandwich; sandwich inserted into back of damp joint.*
3. *New mortar held in position with pointing iron whilst top, then bottom sheet of film withdrawn. Pointing iron then used to compress newly placed mortar; surplus trimmed off, if necessary, with end of sharp, bent knife blade.*

I.McC.

Figure 4.3 Methods for cutting out and filling fine joints

Method 3: pointing gun with a modified nozzle
Several proprietary pointing systems purport to be suitable for the pointing of thin joints. However, a nozzle which can be inserted into a narrow joint of 5 mm and less is not yet available.

During 1986, RTAS undertook some field trials on the pointing of narrow joints 2–4 mm wide. The nozzle of a proprietary mortar gun was adapted by sweating on a piece of sheet copper approximately 2 mm across, 10 mm wide and 25 mm long. Following joint preparation as described above, the gun was filled with a mortar of 2 parts lime putty and 1 part silver sand. The nozzle was inserted in the joint and the mortar was placed, the back of the joint being filled first. Thick wire was used to tamp the mortar when the joint was half full and on completion.

Method 4: taping and pointing (See Figure 4.3)
A strip of adhesive carpet tape is laid over the joint and slit into the open joint with a sharp knife blade. The edges of the tape are pressed into the cut and the mortar pressed home with a pointing key. After compacting as much mortar as possible into the slit the tape is carefully peeled away. The method is suitable for brick or stone.

Important note
It must be accepted that all methods of pointing thin joints will be time-consuming if the job is to be done properly and in a manner which is cost-effective in the medium- and long-term life of a building. The work must be done by tradesmen with a patient disposition and a pride in detailed work.

4.4 NOTES ON MORTAR MIXES

Tables 4.2 and 4.3 show a range of mixes widely used in joint filling, but reference should also be made to the standard recommendations and codes, especially BS 6270: Part 1 1982 'Cleaning and Surface Repair of Buildings'.

The mixes shown in Type A of Table 4.2 are representative of the great majority of historic mortars until the last half of the eighteenth century, used in all conditions and in all types of masonry. Exceptions are some mortars of the Roman period or late mortars following the Roman tradition, which are of the B type (brick dust and natural pozzolanas). Hydraulic lime mortars occur from about 1760 until the 1950s. Portland cement mortars may be found after 1825 but especially after 1856.

Examination and analysis will help determine what the new mortar should be, but the strength and porosity of the masonry units and the exposure of the masonry face are critical factors in the final selection. The schedule at Table 4.3 is an indication only of where some typical mortars might be used.

Hard mortar pointing (flint and granite)

Some tough, very durable masonry materials such as granite, flint and engineering quality brick are traditionally bedded and pointed in very strong mortars.

Table 4.2
Range of recommended mortar mixes for historic building fabric

Type A	*Lime : Sand*	*Type B*	*Lime : Sand : Set-additive*
A1	1 : 4	B1	1 : 4 : $\frac{1}{2}$ brick dust
A2	1 : 3	B2	1 : 3 : $\frac{1}{2}$ brick dust
A3	1 : $2\frac{1}{2}$	B3	1 : 3 : $\frac{1}{10}$ low sulphate PFA
A4	1 : 2	B4	1 : $2\frac{1}{2}$: $\frac{1}{4}$ HTI
A5	1 : 1	B5	1 : 1 : $\frac{1}{4}$ HTI

Type C	*Hydraulic lime : Sand*	*Type E*	*Masonry cement : Sand*
C1	1 : 4	E1	1 : 6
C2	1 : 3	E2	1 : 5
C3	1 : $2\frac{1}{2}$	E3	1 : 4
		E4	1 : 3

Type D: lime mortar gauged with Portland cement

	finished mix	*based on lime : sand*	*by adding cement to coarse stuff*
	Cement : Lime : Sand	Coarse stuff	Cement : Coarse stuff
D1	1 : 1 : 6	1 : 6	1 : 6
D2	1 : 2 : 8	1 : 4	1 : 8
D3	1 : 2 : 10	1 : 5	1 : 10
D4	1 : 3 : 12	1 : 4	1 : 12
D5	$\frac{1}{2}$: 3 : 12	1 : 4	1 : 24
D6	$\frac{1}{4}$: 3 : 12	1 : 4	1 : 48
D7	$\frac{1}{8}$: 3 : 12	1 : 4	1 : 96
D8	$\frac{1}{10}$: 3 : 12	1 : 4	1 : 120

Whilst the masonry units themselves are unlikely to suffer from the use of such mortars the condition of the wall is likely to deteriorate through water penetration. Fine shrinkage and movement cracks almost invariably result from the combination of dense impervious mortar and dense impervious stone or brick. Water tends to 'sheet' on the surface and percolates or is driven into the wall. New mortar should be as soft and permeable as the exposure and loading permits. All the above materials should be pointed in mortars no stronger than D2, D3, D4 and C2 (schedule of typical mortars, Table 4.3). The nature of flint and the way it is used in construction makes the deterioration of even small amounts of mortar on the face a potential structural failure.

The water ratio must be kept as low as possible and the work protected to allow very slow drying. Never carry out flint bedding and pointing in frosty weather.

The weakness of flint walls lies in the small unit size, the vulnerability of the mortar due to the impermeability of the stones, problems of mortar adhesion and problems of bonding flint facings to backings. Frequently the only answer to a bulging wall is to take down and rebuild, if necessary working against a shutter board to bed the flints in 300 mm lifts, backfilling with mortar and pointing after

Table 4.3
Schedule of typical mortars*

Material	*Sheltered*	*Exposed*
Rather weak limestone or brickwork	A1-A2	A3-A4
Very finely jointed brick or stone	A5	A5
Rather weak sandstone	B1-B2	B3-B4
Limestone, or brickwork of moderate durability	C1 A1-A2-A3 B1-B2	C2-C3 D8-D7-D6 D5-D4
Sandstone of moderate durability	B1-B2 C1-C2	C3 D8-D7-D6 D5-D4
Limestone, sandstone or brick of good durability	Any of groups A-B-C	C3 D8-D7-D6 D5-D4-D3 D2

*The letters and numbers refer to mixes listed in Table 4.2.

the shutter is struck. To help overcome poor mortar bonding to the flint, the tails of the stones may be dipped in a cement and sand grout which has been gauged with PVA (polyvinyl acetate). RTAS mortar analysis of old mortar from flint walls has shown that animal fat was frequently included to assist mortar adhesion and water repellence. To achieve back-bonding selected flints may be tied round with heavy-gauge stainless steel wire extended beyond the end of the flint in a loop in the manner of a cavity tie, which is then bedded into the backing. 'Through bonders' of flint, where it is not possible to introduce other stone or brick lacings, can be made up of two flints bedded tail facing tail and joined with a strong epoxy mortar. The tails of these flints must be wire brushed and not grout-coated.

Mortar packing must be very thoroughly carried out because tamping and pointing to depths of 50 mm may be required. Although additives are not to be used in preference to technique, difficult situations may be aided by priming with a 1:10 acrylic emulsion:water mixture.

Granite and engineering brick in severe exposures where there are water penetration problems may require an elaborate pointing procedure designed to include a water barrier. A recommended and tried procedure is to cut out to 50 mm (2 in), push in a backing rod of polystyrene foam, inject a two-part polysulphide

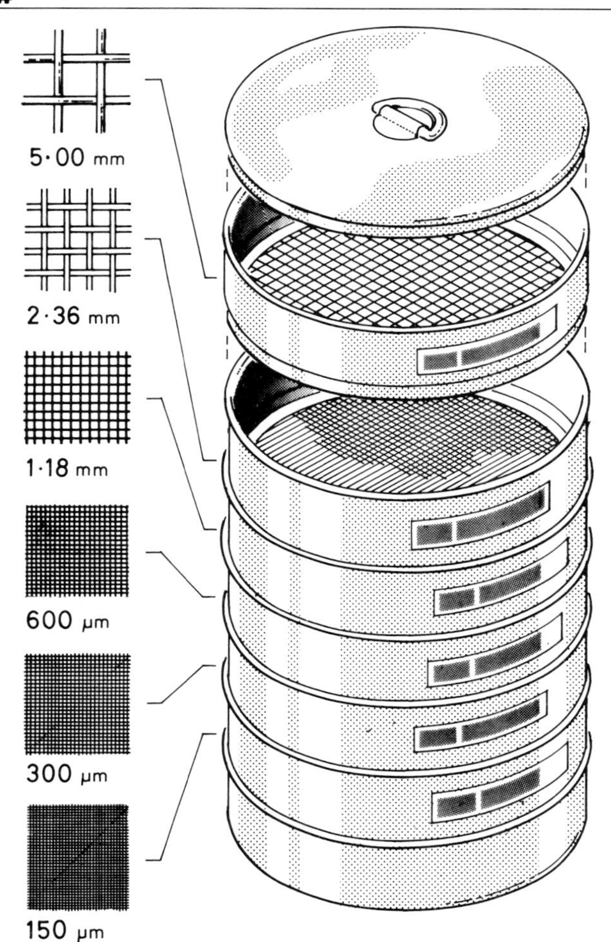

AGGREGATES: GRADING AND BATCHING

Clean, well-graded aggregate is essential to the good performance of mortar or plaster. Small sized aggregate fits between the larger with an even distribution of binder paste. In terms of size, "well-graded" aggregate will range largely between 2·36 mm and 150 microns.

On the left is shown a stack of standard sieves with their relevant mesh sizes. To test the grading of a sample, a known volume of aggregate is placed in the top sieve and the stack is shaken, mechanically or by hand, until the whole sample has been sifted.

When the stack is separated, the amounts which have been retained on each sieve can be assessed as a percentage of the original sample. The bottom tray will contain any material finer than 150 microns. This should never be more than 10% of the total for plasters or more than 15% for mortars

The table below sets out recommended gradings for various functions and should be followed as a general guide, for best results.

BS SIEVE SIZE (to BS 410)	% OF TOTAL SAMPLE PASSING B. S. SIEVES FOR :- MORTARS	EXTERNAL RENDERS	LIME PLASTER	GYPSUM PLASTER
5·00 mm	100 %	100%	100 %	100 %
2·36 mm	90-100 %	90-100 %	100 %	100 %
1·18 mm	70-100 %	70-100 %	90-100 %	90-100 %
600 µm	40-100 %	40-80 %	55-100 %	55-100 %
300 µm	5-70 %	5-40 %	5-50 %	5-50 %
150 µm	0-15 %	0-10 %	0-10 %	0-10 %

I. McC.

Figure 4.4 Grading and batching of aggregates

mastic using a mastic gun and complete the joint with at least 25 mm conventional pointing. The seal may also be formed in a flexible epoxy resin which is more expensive but is likely to have an even longer life. Proprietary water-repellent mortars have also been used successfully on granite, but deep (50 mm) pointing, well packed, is essential.

Damage to joints by masonry or mortar bees

Some wild wasp and bee species will burrow into soft mortar joints and even some weak stones in lieu of their normal habitat of easily eroded rock and earth banks. A common bee of this type is 'osmira rufa'. The holes burrowed by the bees can be up to 75 mm deep. The bee will lay several eggs at the end of its hole, each egg being provided with enough nutrition to enable the development and emergence of an adult bee to occur. Each new bee or wasp will in turn burrow its own hole. (See also BRE Technical Information Leaflet 64, June 1980, 'Damage Caused by Masonry or Mortar Bees'.)

The treatment of a masonry bee problem needs to deal with the unhatched pupae as well as the existing bee population. The wall surface can be made unattractive to bees and wasps by spraying with an insecticide. The pupae problem can be dealt with by raking out and filling the joints, accompanied by hole injection and spraying.

RTAS research at Byland Abbey, West Yorkshire

Byland Abbey, a scheduled ancient monument in West Yorkshire, receives specialist maintenance and conservation work from the directly employed labour force (DEL) of the Yorkshire and Lancashire Area Office of English Heritage. During 1985 and 1986, the soft mortar of several areas of exposed corework was colonized by several species of masonry bees and predator wasps. During 1986 a practical research project was undertaken on the affected mortar, to develop and test a treatment system which would enable the historically valuable mortar to be left intact.

A test area of 2 m^2 was undertaken as follows:

- A clear plastic tube of 6 mm ($\frac{1}{4}$ in) diameter and 100 mm (4 in) length was fitted to the nozzles of two mortar guns
- The first gun was filled with a paste of permethrin and water. The tube was inserted to the rear of each bee hole and a small plug of paste inserted
- The second gun was filled with a mortar of one part lime putty, one part matching sand and a small amount of plasticizer. Each hole was then pointed with a mortar which dried to match the original
- The wall was then sprayed with an insecticide based on bendiocarb

The method of application of the toxic paste and the mortar plug was found to be relatively fast and suitable for two people working in parallel. The long-term results of this work cannot be evaluated yet but the theory is good and RTAS is optimistic. The system tried is far less damaging to historic fabric than re-pointing in hard mortar.

5 THE REPAIR AND MAINTENANCE OF BRICKWORK

5.1 BRICK TYPES

Clay bricks

Bricks are made mainly from sand and clay. Their process of manufacture involves preparation of the earth, moulding, drying and burning of the brick shapes. In general, good clay bricks have a compact texture, are reasonably free from cracks, lime, stones and pebbles and the harder varieties give a metallic ring when struck with a trowel. Good bricks are well and uniformly burnt.

Other types of bricks

Clay bricks are different from *sand-lime (calcium silicate) bricks* in materials and method of manufacture although similarly susceptible to deterioration by damp and salt crystallization. Sand-lime/calcium silicate bricks are made by compressing a mixture of damp sand with 7 per cent of slaked lime at pressures of up to 200 tonnes into precise shapes of high uniformity. Unlike clay bricks they are not burnt but steamed at pressures of 900–1,900 kPa (120–250 psi) for twelve hours in an autoclave. These bricks are smooth, fine-textured and light in colour. The sand-lime brick process was patented in Germany in 1881 as the result of experiments in making artificial stone. Production began there in 1894 but did not reach the UK until ten years later.

Slag bricks are different in materials, manufacture and hence deterioration. Slag bricks are made by running molten slag into iron moulds. The blocks are removed while the interior is still molten, and then annealed in ovens. Slag bricks were used primarily for pavings and are principally nineteenth century in origin.

5.2 CLAY

Components of clay

The main components of the most common clay minerals are silicon dioxide SiO_2 (silica) and aluminium oxide (alumina) Al_2O_3. Clay minerals are composed of a series of wafers, each made up of two three layers of alumina and silica. Water molecules between these wafers cause swelling, and drying conditions cause contraction. The characteristic plasticity of clay is due to the ease with which the thin wafers slide over each other when wet and under slight pressure. Other constituents are limestone, iron, magnesia and sulphates of magnesium, calcium, sodium and potassium, together with organic matter.

Sources of clay

Almost every geological period since the carboniferous has yielded clays suitable for making bricks. All the different kinds of clay rocks are grouped as 'argillaceous' (argilla:white clay). Clay is typically a deposit of a still water bottom such as sheltered estuaries, great depths undisturbed by waves or tides, or lagoons. 'Brick-earth' is the description given to silty clays deposited as wind-blown dust during the last Ice Age. In the UK these deposits are principally found in the south east.

The most common form of compacted clay is *shale*, which has a laminated style of bedding. Water-deposited clay consists, initially, largely of water, which is replaced as the exposed deposit dries out with minute, air-filled pores. Subsequent pressure squeezes out much of the air and accentuates the bedding. If mica flakes are present the shale will be fissile (that is, will split easily into thin layers). *Mudstone and marl* are other types of clay rock with much wider spacing of bedding planes, or without lamination at all. (*Note*: 'marl' may also be used to describe calcareous clay formed at the bottom of pools or shallow lakes and including a high percentage of freshwater shell fragments.)

5.3 FIRING OF BRICKS

The firing process is the key manufacturing step in developing the characteristic properties of brick.

Sandy clays, or mixtures of clay and quartz sand are required for the manufacture of bricks. During firing, water and carbon dioxide are driven off and aluminium silicate (mullite) and more quartz are formed. Sodium or potassium in the clay causes some melting. As the molten clay cools, glass is formed. With high temperatures there is an increase in the formation of glassy material lending greater hardness, density and durability. Primitive firing conditions produce an inconsistency which at best lends attractive and subtle variations in colour and texture, and at worst results in highly distorted or soft underburnt material which will not weather well. Firing temperatures vary according to conditions and fuels, but a range of between 800°C and 1200°C is usual.

Clamps and kilns

The earliest burnt bricks were fired in clamps the form of which varied considerably. The principle was to construct a floor of fired bricks and a level site, with an arrangement of channels running across its length and breadth. These channels or flues were filled with fuel (wood, charcoal, furze and later coal). The clay was prepared either by spreading and treading out on the ground on a layer of straw to remove large aggregate which would cause splitting and disruption when the clay was fired, or, more efficiently, passed through a pug mill with revolving paddles to knead the clay and squeeze out unwanted material. The earliest bricks were cut to size from a thin slab (usually reinforced with short chopped straw), but from early times pressing into moulds was in use as well. The green bricks, with more fuel stacked between them, were built up on the floor of the clamp and spaced to allow the fire to penetrate beneath them. Finally, a thick layer of burnt bricks and clay matrix was spread over the stack and the clamp was fired. A clamp of this kind might burn for several weeks.

Clamp burning preceded kiln burning and is still carried out to a very small extent. The bricks produced are not uniformly burnt and whilst a large proportion is satisfactory, those on the outside are underburnt, and those near the fire-holes and in the heart of the kiln may be misshapen and cracked owing to the excessive heat.

From the Roman period in Britain *up-draught kilns* are to be found coexisting with clamp burning. Up-draught kilns consist of two brick chambers one stacked with bricks over another filled with fuel. The brick chambers, bonded with clay were frequently constructed into the slope of a hill for economy of excavation and construction and to facilitate the loading and unloading of the upper chamber. This kiln had firing temperatures of about 1000°C.

Another traditional system of firing brick, which also survives, was the *down-draught kiln*. These are comprised of a chamber with straight walls and a vaulted or domed top, lined with firebricks. Fireboxes are formed at intervals at the base of the walls. The heat from these passes between the walls and continuous screen walls constructed nearby, upwards to the arch which deflects it down through the honeycomb stack of green bricks, the gases escaping through perforations in the floor to a horizontal flue connected to a tall chimney. The heat control possible in this type of kiln enables it to produce high-quality bricks. The large and regular outputs of bricks today are produced mainly in continuous kilns where the operations are uninterrupted, each chamber in turn being loaded, dried, burnt, cooled and emptied; the waste heat is utilized to dry and pre-heat the green bricks.

Colours of bricks

The colour of brick is affected by chemical constituents, by temperature, by atmospheric conditions, and by sand rolling.

Most bricks turn red in an oxidizing atmosphere at 900°C to 1000°C. Above this temperature, the colour may change to plum or purple, or to brown or grey at 1200°C. In a reducing atmosphere, in which the supply of oxygen is restricted or cut off, purple or blue bricks often with black cores are produced. The following constituents have the following colour effects:

Constituent	Colour
Lime (high percentage) with iron traces	*White*
Lime (low percentage) with iron traces	*Grey*
Chalk (low percentage) with iron traces	*Cream*

Iron oxide (up to 2%)	*Buff*
" " at 900°C in a reducing atmosphere	*Brown*
" " at 900°C in an oxidizing atmosphere	*Bright salmon*
" " at 1100°C in a reducing atmosphere	*Red*
" " (7%–10%)	*Blue*
" " " " + manganese oxides	*Black*

The presence of vegetable matter will also produce black colouration, sealing carbon in a black core, especially when there is a sudden increase in temperature. A method of deliberately producing black bricks is to pour coal tar in the fire boxes of a down-draught kiln just before burning is complete.

Surface colour may be produced by adding finely ground metallic oxides to sand sprinkled on the bricks before burning, for example, manganese – brown, chromium – pink, cobalt and manganese – black, antimony – yellow, copper – green, cobalt – blue.

5.4 CAUSES AND EFFECTS OF DETERIORATION

Manufacture

Cracking and warping of bricks creating incipient weaknesses may be due to exposure of the green bricks to direct sunlight or rapid-drying winds when placed in the drying sheds. Similar effects will be produced by placing green bricks too soon into the kiln. Hard, warped bricks may also be produced by too low a sand content. Too much sand produces bricks which are brittle. Bricks which have a high lime content, especially when the lime constituent is in large pieces, will be liable to swell and burst after firing when the brick is exposed to moisture, due to the lime slaking. Bricks with a high percentage of salts present will exhibit unsightly surface scum after firing and will tend to attract moisture. Slop moulding, the process of dipping brick moulds into water each time of using to prevent the clay sticking to them may also lead to surface cracking and poor shape. Sand moulding generally produces better bricks with sharper profiles.

A single brick which shows signs of deterioration in an otherwise sound expanse of brickwork will have peculiar individual deficiences which allow this. Such bricks may not have been fully fired or may have been made from clay containing impurities as described above. While some brick types may be of good quality they may have low durability and in aggressive environments will deteriorate relatively quickly. Rubbing bricks are of this type.

Detailing and weathering

The penetration of brickwork by water is one of the commonest and potentially one of the most damaging failures that can occur. Water can percolate and progress by capillarity within a wall from entry points such as defective copings and flashings and broken down damp-proof membranes. It can also penetrate through the bricks and joints, especially on features such as ornamental strings and cornices. Bricks differ in their tolerance of wet conditions according to their properties.

Zones of a building where bricks are most liable to deteriorate are chimney stacks, parapets, aprons under windows, quoins and plinths. These areas often receive large amounts of water and saturation of the brickwork is a key to its breakdown. By contrast, further areas which are liable to deteriorate earlier are surfaces which are not washed at all, such as courses under projecting strings, where there is a slow build up of dirt and where water in fine droplet form will tend to remain on the surface. The formation of destructive salts will almost certainly occur as a result.

Soluble salts

If crystallization of soluble salts occurs within the brick, some spalling or disintegration of the surface will result. The salts may be derived from the clay itself or from the mortar, the run-off from materials such as limestone, or polluted or marine environments. A major source of soluble salts is the soil from which salts can be carried up by rising damp. In this case the salts crystallize just below the surface of the brickwork within the evaporation zone at the top of the damp section of wall.

Sulphate attack on historic brickwork is generally confined to chimney stacks and especially on industrial monuments such as kilns.

Brickwork is very liable to deterioration where there is direct washing from limestone strings, band courses and cornices. This leads to disruption of the surface due to an accumulation of calcium carbonate and calcium sulphate filling the pores of the bricks.

Damage from movement

Brickwork which undergoes movement may experience cracking, surface defects and displacements of varying types and degrees. Principal causes are often differential settlement, changes in the nature of the soil for various reasons and excessive external forces. Initially the structural stability of a wall and the continuation or otherwise of movement need to be determined.

The decision whether or not to repair cracks and cut out bricks will depend on whether there is any risk to the stability of a wall and – if not repaired – whether the cracks are likely to encourage rain penetration. Fine cracks are not very conspicuous and pointing can often make them more unsightly. However, water may penetrate them and their repair may be advisable. Wider cracks will generally need to be repaired.

Special attention is required for horizontal displacement where a course of bricks projects so as to form a ledge where water may collect. Cracks due to differential settlement may involve distortion of the brick and movement inwards and outwards so that the faces are not in the same place. This situation will usually mean at least some cutting out or taking down and resetting.

Frost

Clay bricks vary widely in their resistance to frost. Highly porous bricks with a large percentage of open pores have poor resistance to freezing. Water penetrates the major pores of the brick, expands on freezing causing the surface to break up and spall. Disintegration of the mortar may also occur. The position and exposure of brickwork are also major determinants of this form of deterioration as frost will damage brickwork only when it is very wet. Common areas of frost failure are free-standing walls, parapets and retaining walls and occasionally brickwork in contact with very damp ground. In some cases, frost may only have accelerated damage due primarily to some other cause.

Fire

The effects of fire vary according to the severity of the fire, the thickness of the walls and the type of brick. Having been fired at temperatures higher than those which normally arise in building fires, clay bricks usually provide excellent resistance to fire. In a severe fire, cracking and bulging of the brickwork is to be expected, in addition to surface damage such as spalling. Some vitrification of clay bricks may occur if the fire is very intense. The sudden quenching of brickwork with water in the course of fire-fighting may cause spalling but generally it does not affect the strength and stability of the wall seriously. More important is the distortion caused by thermal movement of members built into the wall which may displace the brickwork and cause cracking. Smoke staining of brickwork may need to be cleaned (see Volume 1, Chapter 5, 'Masonry cleaning').

5.5 REMEDIAL WORKS TO BRICKWORK

Remedial works to brickwork should be carefully selected on the basis of a thorough analysis of the causes of the evident deterioration and should always be kept to a minimum. This section considers several techniques which are appropriate to the repair of particular problems.

Replacing whole bricks

Surface repair of brickwork may involve either the replacement of individual bricks or the replacement of a defined area of brickwork. The choice of repair method will often depend on the ease with which cutting out may be achieved and this in turn will relate to the brickwork bond used and the strength of the mortar. Whatever the repair method, the cutting out operation should be that which least disturbs the sound brickwork.

Replacement bricks should match the size, colour, texture and as far as can be ascertained, the durability of the surviving existing brickwork. The problem with replacing historic bricks is that they are almost always non-standard sizes and frequently non-standard colours. The sizes of bricks obtained for repair should be checked carefully against the dimensions of the original bricks, especially where new bricks are to be used, as there may be enough difference to create problems in coursing, and bonding in replacement bricks with existing work. In some cases, it may be necessary to have new bricks specially made for the work. There are a number of good suppliers available in the UK.

Cutting out decayed bricks and rebedding matched specials or second-hand bricks requires skill and experience if the new work is not to be obvious. This seventeenth-century wall requires relatively small numbers of bricks to be replaced, but they have been fired for the job because of the unusual size. Great care is taken to maintain the gauge, irregularities of bond and character of the mortar. The mortar being used here is based on one part of hydraulic lime mixed with two parts sharp, gritty sand and two parts of a soft sand providing the final mix with a creamy colour.

Second-hand bricks

Occasionally suitable second-hand bricks may be located through local observation and inquiry. A careful inspection must be made of the condition. It would be unwise, for instance, to select from a second-hand source bricks which had not weathered externally before or had been badly damaged in demolition work or handling. Where supplies are very limited it may be necessary to cut single bricks with a diamond wheel to produce stretcher or header slips.

Reversing bricks
Occasionally a damaged or extensively weathered brick may be reversed by carefully cutting it out, cleaning off the mortar and placing it in reverse position back in the wall. There is little point in reversing decayed bricks of poor quality since the same decay process will take place in the other half of the brick once exposed to the weather. The cutting out and reversal of bricks is labour-intensive and in practice is rarely carried out. Hydrochloric acid may assist the breakdown of hard mortar adhering to the brick beds during the cleaning process; bricks must be soaked before acid treatment in clean water and thoroughly washed afterwards.

Making new 'specials' to match
Inquiries should be made through local sources, the Brick Development Association or a conservation officer of the local or county authority who may know of the several small brickworks which are producing 'specials' as matches for medieval and later types of brick. An accurate assessment will need to be made of the numbers of bricks required and over what period they are likely to be needed. It is always sensible to over-order generously because more bricks than are estimated are nearly always required and it is useful to have a small stockpile for later repairs. Specials can very rarely be produced quickly and advanced planning is essential.

The building as a source
A last alternative source of replacement bricks which may on rare occasions need to be considered is the building itself. It may be necessary to remove bricks from another place already extensively altered, from collapsed boundary walls or from neglected outbuildings. Of course the removal of such bricks requires a lot of thought and discretion and in no circumstances should a valuable building be robbed for materials or evidence of archaeological significance destroyed.

Toning in replacement bricks
Where replacement bricks have been inserted or where new bricks are juxtaposed against old in a rebuild or an extension it is sometimes thought desirable to reduce the bright appearance of the new work by the application of a wash. In general, it is better to leave natural weathering and soiling processes to blend in the new work, but if necessary a traditional treatment of soot wash prepared by leaving a bag of soot in a bucket of water overnight or a diluted paint stain may be used. These techniques must always be carried out on an unobtrusive sample and left to dry out before an assessment is made and before a large area is treated.

Brick slips
When cutting out is likely to seriously disturb the adjacent sound brickwork, consideration may need to be given to the insertion of a brick slip rather than a full-size unit. Brick slips are facings of about 25 mm (1 in) thickness although it may be possible to cut thinner slips from whole bricks.

Repair using brick slips should be limited to individual bricks or to relatively

An excellent, recent repair of diaper-patterned sixteenth-century brickwork. Re-pointing is kept well back from the weathered arrises, new bricks have been specially made to the correct size, colour and texture and no attempt has been made to correct or add to the pattern as found.

small areas of brickwork. The slips must be solidly bedded in the prepared indent on a bed of mortar with or without the assistance of purpose-made or patent clips. The cavities formed should be clean and regular and the mortar used compatible with that in the original brickwork. In particular the refacing of larger areas with slips bedded in an epoxy or other resin mortar is to be discouraged because of the membrane effect which will trap moisture and may result in spalling of the patched area or adjacent bricks. Where the area requiring repair is more extensive, it may become necessary to tie in a new leaf of matching brickwork, using header bricks, suitable ties or anchors, as appropriate.

Repairing gauged and decorative brickwork

The repair of ornamental brickwork raises particular problems. The greatest difficulties are met when repairing rubbed and gauged brickwork. Decorative work formed of cut or moulded ordinary bricks and patterned and polychrome brickwork need careful handling if the repairs are not to mar permanently the intended ornamental effect. Specials for replacement are rarely available from stock and contact should be made as early as possible with potential suppliers.

Rubbed and gauged brickwork

Traditionally, soft bricks, cut to shape and rubbed to a high degree of accuracy, were laid in beds of lime putty, the joints rarely exceeding 3 mm in thickness. Repairs to rubbed and gauged brickwork should only be carried out in the most pressing circumstances, for example, where collapse of an arch is imminent or where there is water penetration.

It is easy to spoil the appearance of such brickwork by attempting to repoint the joints, and raking out joints will inevitably damage the surfaces and arrises, thus destroying the sharpness and precision which are vital to the effect of the finished work.

Where repairs are essential, it is best to dismantle the rubbed brickwork carefully and rebuild it, using the original bricks where possible and lime putty as a jointing medium. Where one or two bricks have slipped it is sometimes possible, after using a purpose-made or hacksaw blade to remove the remaining mortar and to ease the bricks back into position, wedging them with a sliver of lead or slate. A 50 cc hypodermic is a useful way of introducing new mortar which should be composed of a slurry of lime putty and finely powdered HTI (refractory brick powder – high temperature insulation) powder. The surface of the joints may be sealed with two brush applications of latex paint which can be gently pulled away after the hydraulic set has taken place. The needle for grouting can be pushed through the latex skin. (See also 'adhesive tape method' Chapter 4, p. 40.)

Patterned brickwork

The repair of patterned and polychrome brickwork creates special problems in obtaining suitable replacement bricks. Where it is not possible to obtain matching bricks, alternative methods of reproducing the decorative work may have to be explored such as modifying the colour of available bricks or plastic repair.

Plastic repair

The use of 'plastic' or mortar repairs is frequently resorted to on grounds of economy and there are historic precedents for its use. In the past badly deteriorated brickwork was frequently patched with coloured mortar and colour washed overall; joints were often ruled in and tuck pointed in an attempt to restore the appearance of good brickwork. Although mortar repairs of damaged bricks may be justified where only small amounts of work are needed the use of this technique is to be deprecated. Very few mortar repairs are of good quality and they can be very disfiguring and in some cases damaging to the bricks they are trying to repair. Matches are often attempted using powdered brick or pigments and lime and cement.

Some of the best mortar repairs for brickwork have been carried out with sharp sand with strong natural colour and with a masonry cement binder sometimes using a SBR (styrene butiadene rubber) additive. Pigments should be avoided and it should be noted that lime frequently imparts an undesirable pastel appearance. A typical mix would be 1 part masonry cement:5 or 6 parts sharp sand/brick dust. Bonding additives should be the exception, rather than the rule.

The cavity to take the repair should be carefully cut out to a minimum depth 25 mm (1 in). Each brick should be repaired individually and the cavities well wetted before the mortar is placed to avoid dewatering. The mortar must be very firmly compacted in layers not exceeding 9 mm thick and is best built up proud of the required finish line and then scraped back with a fine saw blade. Pointing of adjacent joints should be carried out as a separate operation.

Treatment of bulges and fractures

In common with other types of masonry a careful diagnosis must be made of the cause of fracturing or bulging. Fracturing is commonly the result of local subsidence or the failure of timber lintels and bond timbers or alterations and additions to the original structure. Bulging may similarly relate to alterations in loadings and frequently occurs due to lack of bond between brick skins or straight joints at wall junctions.

On the basis of this diagnosis a decision must be made to:

- Leave alone
- Cut out and point
- Stitch and grout in situ
- Take down and rebuild

Leave alone

In this situation the cause of fracturing will have long passed and there is no problem affecting stability or water penetration. More damage may be caused by cutting out or attempting to grout than by doing nothing.

Cut out and point

The decision to cut out and point a fracture would be made where the cause of fracturing had passed but the fracturing remained as a source of water penetration and potential structural weakness.

The brick quoin of this seventeenth-century building has torn under load and slight, unequal settlement. Although the movement appears to have ceased, cracks up to 3 mm have been created. Where bricks have fractured completely they should be carefully cut out, cleaned and jointed in clear epoxy resin adhesive and re-set in position in lime mortar to match the original. Fractured joints should be cut out, flushed, tamped and repointed. If old bricks cannot be jointed, second-hand or new specials to match the existing size exactly must be used to stitch across the fracture. Much of this wall could be re-pointed to advantage by removing the nineteenth-century mortar with its penny-round joint pattern and lost pigment and copying the original.

The fracture should be carefully cut out with hacksaw blades, masonry saws, plugging chisels and/or diamond wheels, flushed with water and pointed. If fracturing has extended through some of the bricks these should be cut out and replaced.

Stitch and grout in situ

The purpose of stitching is to prevent further movement taking place which would threaten the stability of the structure. It is often introduced into walls where the situation is in some doubt, sometimes unwisely. The decision to stitch and grout in situ would be made where fracturing extends through bricks and joints and extensive cutting out for the full height of the fracture may be necessary. This is often found, for instance, in a chimney stack with large or multiple flues. These fractures will usually be cut out in metre lifts, the fractured bricks replaced with whole ones and hand grouting and deep tamping of fractured joints or rubble fills carried out.

If the fracturing has resulted in a two-way movement it may be necessary to introduce concrete stitching at intervals behind the brick facing. These may be precast dovetail sections or may be cast in situ. In both cases the facing bricks are removed for a depth of, say, three courses and the opening extended 450 mm (18 in) either side of the fracture. After the stitch is placed the bricks are replaced exactly in the former position.

Other forms of stitching include the introduction of brick reinforcement or stainless steel mesh or wire set into new mortar joints. Special ties may be introduced from an internal face to bond two skins together. Some of these systems rely on an epoxy mortar to anchor the ties and are forms of resin anchor (see also *BRE Digest 257* 'Installation of Wall Ties in Existing Construction'). Others lock two elements together such as a patented 'remedial fixing', consisting of a stainless steel or aluminium bronze tie with an expanding head at each end. These ties are set in an 8 mm drilling and the expansion takes place by using a locking key.

Take down and rebuild

In most cases where walls are bulging or leaning and when there is serious displacement around a fracture the proper course of action is to carefully take down and rebuild, salvaging all possible material. The rebuild should follow the original coursing, bonding and joint profile. This is an extreme measure which is the least desirable in conservation terms. Conscientious recording before taking down is absolutely imperative. Tough polyethylene sheets can be pinned to the face to trace typical bond and joint patterns. Cut or moulded bricks must be numbered on the bed as they are taken down. Bricks must be stored on dry bases and protected from abrasion and saturation.

Treatment of joints

A great deal of unnecessary pointing of brickwork is carried out. A rough weathered face of lime mortar which can be damaged with a knife blade is not an indication that new mortar is required. Because the mortar joints form such a

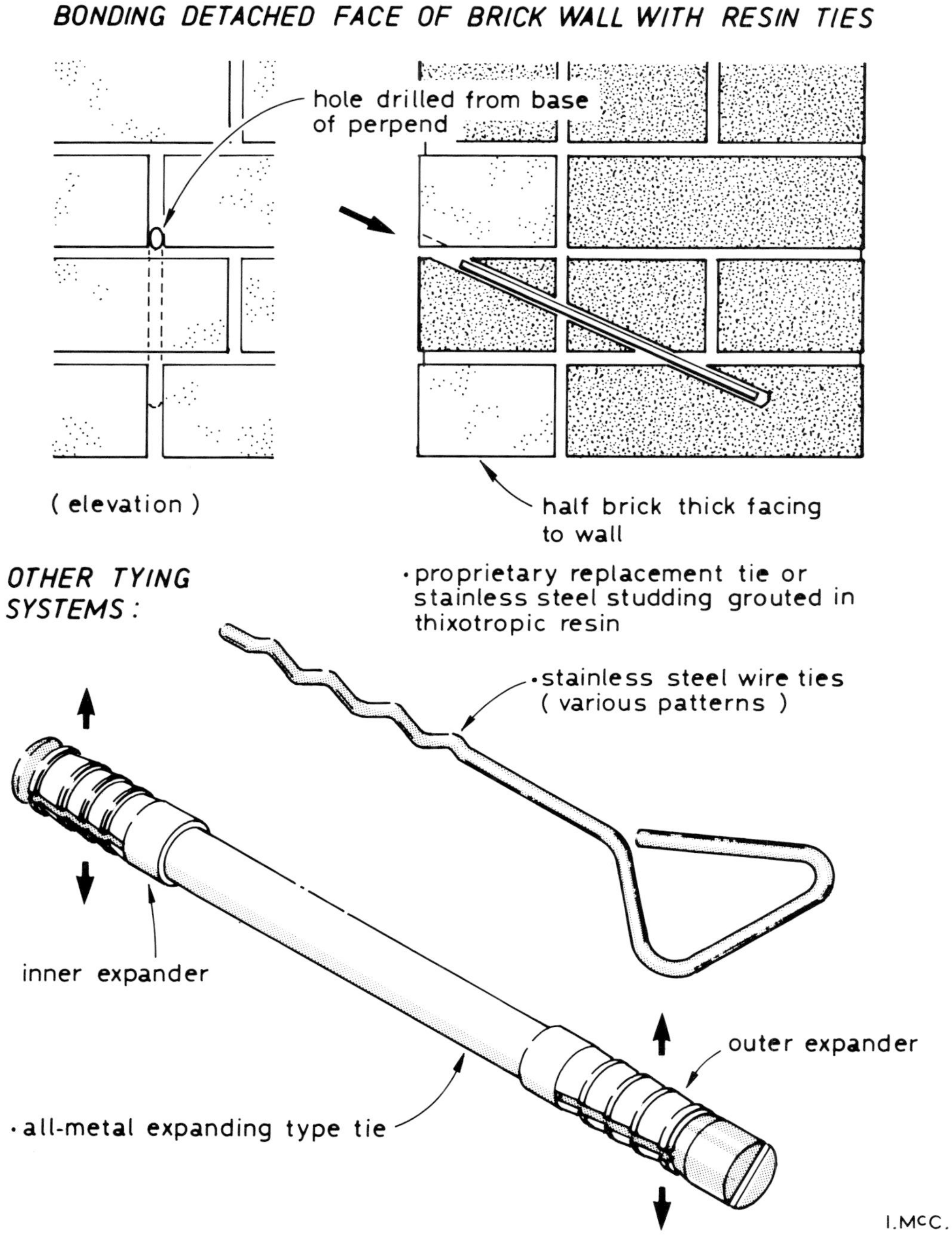

BONDING DETACHED FACE OF BRICK WALL WITH RESIN TIES

hole drilled from base of perpend

(elevation)

half brick thick facing to wall

OTHER TYING SYSTEMS :

• proprietary replacement tie or stainless steel studding grouted in thixotropic resin

• stainless steel wire ties (various patterns)

inner expander

outer expander

• all-metal expanding type tie

I.McC.

Figure 5.1 Methods of tying brick skins together

English bonded $2\frac{1}{4}$in bricks with wide lime mortar joints in weathered but perfect condition. Because this mortar could be broken up by driving a large screwdriver into it, it was considered by a surveyor to need raking out and repointing. Familiarity with the weathered appearance and long life potential of traditional materials is essential if this kind of destructive activity is to be avoided.

large percentage of the surface area of brickwork, pointing or repointing is a major decision which can drastically affect the appearance and condition of the whole wall. Only when joints are truly friable and have weathered back extensively and water retention is encouraged should pointing be embarked on. It may also be necessary where a dense impermeable mortar has been introduced in earlier remedial work and is causing sacrificial deterioration of the adjacent bricks.

Satisfactory mortars for old brickwork are usually quite weak. Lime:sand mortars in the standard proportions of 1:3 are often satisfactory with minor additions of set additives. Hydraulic lime and cement:lime compos are necessary where the exposure is more demanding but dense cement-rich mortars should always be avoided even when the bricks themselves are dense and hard.

Careful record must be made of any evidence of treatment of the original joint face. Joints in old brickwork were normally finished flush but may have been struck or double-struck ('birds-beak'), or ruled with a trowel-edge or jointing tool ('joints jointed'), or ruled with a coin edge ('penny-round'), or coloured and false

This interesting wall has had problems for many years because of the high proportion of rather poor quality bricks used in its construction. To overcome the problem, large areas have been slurried with coloured mortar onto which false, black joints have been drawn. Elsewhere, black ash pointing has spread over the weathered arrises. Much of the face is conservable by a mixed technique of replacement, pointing and tuck pointing.

jointed ('tuck' or 'bastard-tuck' joints). These finishes should be repeated. If the edges of the bricks have become rounded with weathering the mortar must be kept back from the face to avoid increasing the width of the joint. If there is no evidence of the original finish it is a good idea to match the surviving weathered mortar by stiff brush stippling after the initial set. Distinctive 'modern' forms of pointing such as 'weather-struck' or 'bucket-handle' must be avoided, however sure the bricklayer may be about their merits.

For detailed information on the pointing of brickwork reference should be made to Chapter 4, 'Pointing stone and brick'.

5.6 CLEANING AND MAINTENANCE

(See also Volume 1, Chapter 5, 'Masonry cleaning'; and Volume 1, Chapter 2, 'Control of organic growth'.)

General considerations for cleaning brickwork, including the preparation of buildings are described in the above note. Although much brickwork can be cleaned very satisfactorily and brick buildings can be greatly enhanced by cleaning, it is also unfortunately true to say that a great deal of damage has been done by careless execution and inappropriate methods.

Washing

Nebulous spray washing will remove dirt effectively, but heavy soil deposits may require an undesirable amount of saturation. Long periods of soaking lead almost inevitably to disfiguring efflorescences. Whilst these can be dry-brushed from the wall there is an increased risk of reappearance following washing. A further hazard is vigorous brushing. Although brushes should be non-ferrous soft wire, or bristle, and never steel wire, even these can scratch and scour the surface of soft facings, especially gauged brickwork ('rubbing bricks'). Nail brushes of compact natural bristle are the most suitable for rubbing bricks, and bristle or phosphor bronze wire for other facings. Saturation can be reduced by using intermittent or 'pulse' washing, which places nebulous sprays under the control of a pre-set clock. Dirt deposits can be removed by relatively short bursts of water at intervals spaced to achieve a progressive softening without soaking.

Glazed brick and semi-engineering or engineering bricks may be cleaned using low-volume medium-pressure water lances and a neutral pH soap, or on small areas, by water, bristle brush and soap, rinsing off carefully afterwards. Water at pressures which damage the brick surface or the mortar joints must never be used.

Mechanical cleaning

Spinning dirt off with a carborundum head removes the surface of the brick and risks 'dishing' and scouring, especially of arrises. Sand or grit blasting commonly tears and pits the cleaned face, greatly increasing the surface area and encouraging re-soiling and an accelerated rate of decay. Although the damage caused by these mechanical systems is usually obvious, the use of at least one of them

Typical damage caused by the use of compressed air and abrasive on the face of nineteenth-century brickwork. The nature of these bricks is too inconsistent to be cleaned satisfactorily in this way and a pitted texture which will soil and weather more rapidly has resulted. This brickwork should have been cleaned with very dilute hydrofluoric acid after pre-wetting.

persists, especially when paint or limewash is to be removed. Small-scale air abrasive tools can be used successfully, but success is so unusual that it is generally safest to recommend against it. Safe cleaning of brick by compressed air and abrasive tools requires much experience, is very slow and therefore expensive. Paint should be removed using non-caustic solvent packs wherever possible.

Chemical cleaning

Some of the best cleaning of brickwork is carried out using weak solutions of hydrofluoric acid. 'Weak' means 2–8 per cent concentration. The soiled area must be pre-wetted to limit the activity of the acid to the surface. This is most conveniently carried out using a water lance. The acid should be in the form of a proprietary cleaner of known concentration which can be used direct or further diluted on site, but dilution of industrial concentration acid must not be carried out on site. Windows, paint and polished surfaces must be fully protected, the scaffold capped and operatives must wear full protective clothing, paying particular attention to face and hands. Careful brush application of the acid should

Typical disfigurement of dense semi-engineering brickwork due to inadequate washing off of hydrofluoric acid after cleaning. The illustration shows what takes place when very weak acid lodges in the weathered joints after washing off. As it trickles from the joint it increases in concentration as water is lost and forms a pattern of dribbles and streaks of colloidal silica which is almost impossible to remove.

be followed by a contact ('dwell') time of about five minutes and thorough washing off with a water lance. A proprietary acid formulation incorporating a rust inhibitor is recommended to reduce the risk of brown staining of mortar when some aggregates are attacked, although this is a fairly unusual problem, and it is normally wiser not to use a 'rust-inhibitor', as white films of calcium phosphate can be left behind. Unless washing off is thorough, a white silica 'bloom' will be left on the bricks from any hydrofluoric acid-based cleaner.

Mortar and mortar slurry stains can be removed using 10 per cent concentration hydrochloric acid or, if stubborn, under an EDTA poultice over a period of days (see Volume 1, Chapter 7, 'Cleaning marble'). Most paints (including graffitti applications) can be softened using a methylene chloride paint stripper applied as a thick paste in attapulgite clay, covered with thin plastic film or a proprietary paste incorporating sodium hydroxide (caustic soda). There is a risk of efflorescence with any brick cleaning agent other than hydrofluoric acid and all must be applied to a properly damp surface and thoroughly washed off.

Surface treatments

Surface treatments applied to brickwork are usually water repellents or graffiti barriers. These are described in Volume 1, Chapter 10, 'Colourless water-repellent treatments' and Volume 1, Chapter 5, 'Masonry cleaning'. These treatments should not be applied to decaying brickwork and the cause of dampness must always be ascertained before resorting to application (refer to Chapter 1, 'Control of damp in buildings').

Silicone water repellents can be useful in assisting brickwork to be partially self-cleansing. This may be important where there is a history of staining from limestone dressings. After removing the deposits the bricks (not the decaying stone) may be treated with a silicone water repellent of the appropriate class. Future deposits tend to be removed by rain. A more positive approach, where possible, is to modify details by inserting a lead or bronze strip into the bottom bed joint of the limestone to throw off potentially harmful and disfiguring deposits.

Consolidants

Very little work has yet been carried out on the use of consolidants in fired clay. Some success has been achieved, however, in preserving external areas of medieval tile by flood application of an uncatalysed alkoxy-silane, and in slowing the deterioration of structural and decorative brickwork by the use of a catalysed and uncatalysed alkoxy-silane system. These treatments are predictably more effective on soft, rather absorbent material such as weathered tiles or rubbing bricks. Hard bricks, even when decaying, do not readily absorb even these low-viscosity consolidants.

REFERENCES

1 Building Research Establishment:
BRE Digest 164: Clay Brickwork 1, HMSO, Garston
BRE Digest 165: Clay Brickwork 2, HMSO, Garston
BRE Digest 257: Installation of Wall Ties in Existing Construction, HMSO, Garston
BRE Digest 139: Control of Lichens, Moulds and Similar Growths, HMSO, Garston.
2 Brick Development Association, *Building Note No 2, Cleaning of Brickwork.*
3 British Standards Institution:
BS 3921: 1974 *Clay Bricks and Blocks*
BS 6270: *Code of Practice for Cleaning and Surface Repair of Buildings*
Part 1: 1982 *Natural stone, Cast Stone and Calcium Silicate Brick Masonry*
BS 187: 1969 *Calcium Silicate Bricks.*
4 Brunskill, R and Clifton-Taylor, A, *English Brickwork*, Ward Lock.
5 Davey, N, *A History of Building Materials*, 1961.
6 Lloyd, Nathaniel, *A History of English Brickwork*, 1925 (covers up to 1800).
7 Bidwell, T, *Repair of Brickwork*, Brick Development Association.

See also Technical Bibliography, Volume 5.

6 REPAIR AND MAINTENANCE OF TERRACOTTA AND FAIENCE

6.1 DEFINITIONS AND INTRODUCTION

This chapter deals with the deterioration, repair and maintenance of the ceramic building materials terracotta and faience, with some reference to Coade stone.

Terracotta and faience: definitions

Terracotta and faience are moulded clay products made from fine, pure clays mixed with other materials such as sand and pulverized fired clay. They are usually well vitrified and have a hardness, compactness and sharpness of detail not normally obtained with brick. They were used widely in the building industry in the period 1840 to 1910.

Terracotta and faience are differentiated principally by their constructional function.

Terracotta is moulded block used in a structural or semi-structural context. In manufacture the clay is invariably hand-pressed into absorbent moulds to form hollow boxes within which there may be clay walls called 'webs' or 'straps' which support the form prior to firing and allow the heat to reach the greatest surface area of clay. Blocks were usually dowelled, cramped and anchored to a substrate or frame by means of iron or steel fixings. The voids at the backs of the blocks were designed to accommodate these fixings which were locked in place by tightly packing the hollows with materials such as concrete, concrete with aggregate or brick pieces, and breeze concrete.

Faience is large solid masonry slabs (tiles) which are fixed as a cladding by bedding in a screed of concrete. They often had a ribbed back, were approximately 300 mm × 400 mm (12 in × 16 in) in size with a nominal thickness of 25–50 mm (1–2 in). These units were often used internally.

In the British ceramics trade, the term 'terracotta' often refers to both structural and non-structural units whose colour and finish is that of the clay from which they are made. 'Faience' often refers to glazed units.

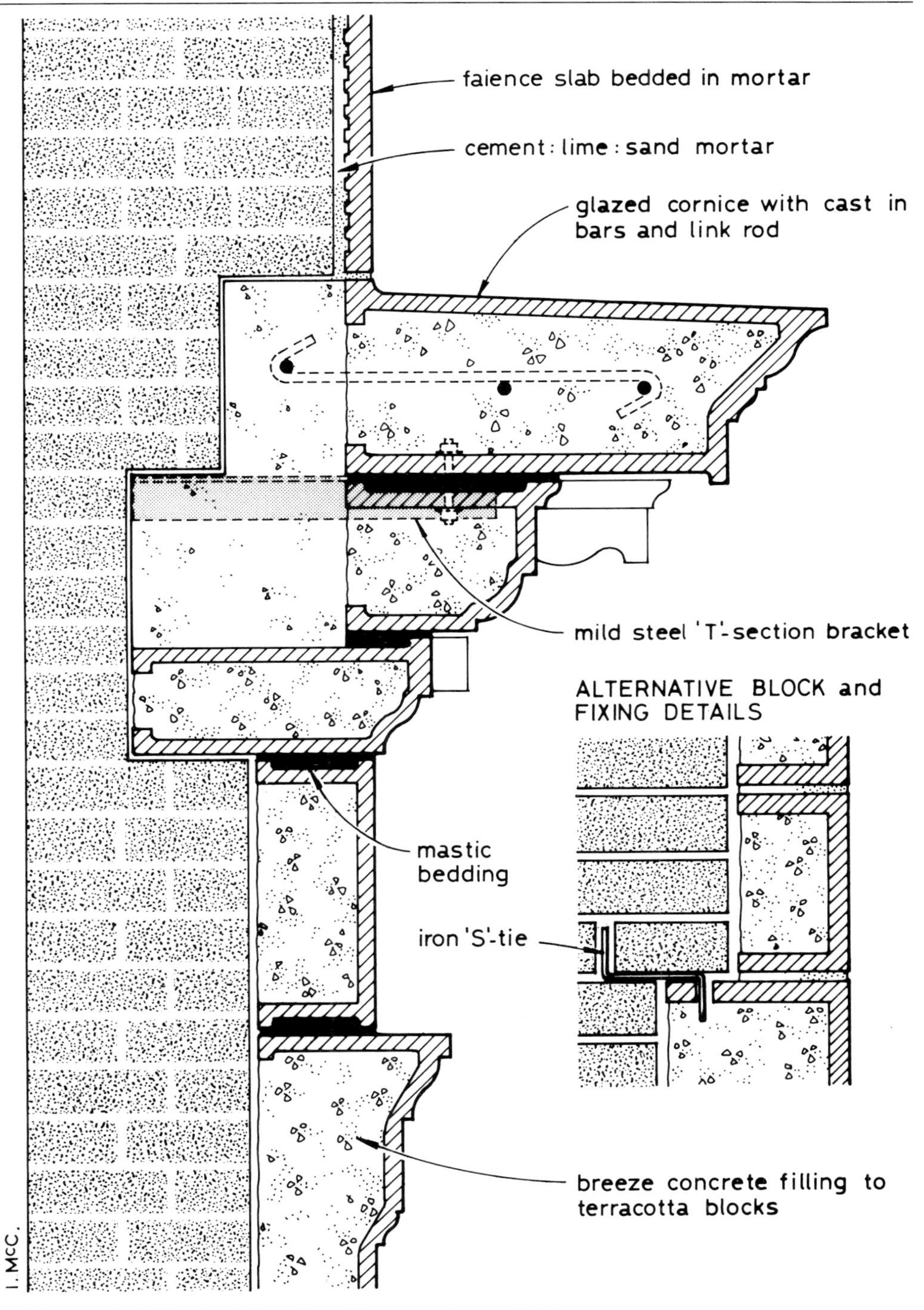

Figure 6.1 Typical details of terracotta and faience

Both terracotta and faience can be plain or decorative, glazed or unglazed. Their clays vary widely in colour ranging from red and brown to dark gray and off-white. There are no British Standards for the materials, although BS 1181: 1971 'Clay Fire Linings and Flue Terminals' makes reference to clays, glazes, materials and workmanship.

History of terracotta and faience

Terracotta is one of the oldest building materials known to man and was used in Babylonia as long ago as 1400 BC. In England the craft of terracotta manufacture dates from the period of Roman occupation and although little survives from intervening years terracotta appears in some notable buildings of the sixteenth century such as Nonsuch Palace, Sutton Place and Hampton Court.

In the early part of the nineteenth century terracotta became popular and in the second half of that century was used to construct large, prominent buildings

Although most of the surviving terracotta in the UK is nineteenth- and twentieth-century. There are some important early examples such as the sixteenth-century detail from Sutton Place shown in this illustration. Because of the high intrinsic value of the blocks, every attempt must be made to retain as much as possible, rather than replace. Mortar "dentistry" repairs, alkoxysilane consolidation of exposed underbody, removal of iron and washing and waxing of detail are all likely to feature in the conservation programme.

such as the Henry Cole Wing of the Victoria and Albert Museum (1866), the Royal Albert Hall (1867–81) and the Natural History Museum (1880–1905). The principal claims made for terracotta were its cheapness, attractive range of colours and textures, plasticity and ease of moulding, but above all, its durability and low maintenance requirements. A timely boost to its use was the stonemasons' strike of 1877 when it provided a convenient substitute for stone.

Doultons were leading manufacturers and suppliers. In the 1820s John Doulton had begun to produce terracotta chimney pots, ridge tiles and garden ornaments. A decade later brother Henry Doulton joined the business and by the late 1870s Doulton was the leading manufacturer supplying terracotta for major buildings.

Terracotta and faience were very suited to the architectural styles of the twentieth century but their growth was hampered by World War One and almost terminated by World War Two. A limited number of manufacturers of terracotta and faience exist today.

Manufacture of terracotta and faience

The several clays and additives used in the manufacture of *terracotta and faience* were selected to overcome as much of the uncertain shrinkage of clay as possible. A shrinkage of $\frac{1}{10}$ to $\frac{1}{12}$ of the lineal dimension of a unit was usual; this had to be known accurately and the mould would be constructed oversize to compensate for this. The clay mixture was ground to a fine powder, thrown into water, finely strained, pugged and kneaded. It was then packed into dry plaster of paris moulds smeared with soft soap. As the plaster drew moisture the packed clay shrank, thus releasing itself from the mould in the space of two or three days at room temperature. The body was then turned out for finishing or tooling. The blocks were dried in a controlled atmosphere. Dependent on the type of glaze and texture required the glaze was brushed or sprayed on to the weather faces of the clay body, either at this stage, before a once-only firing, or after an initial firing to the biscuit condition before refiring to vitrify the coating.

Terracotta blocks were usually 1 to 3 cubic feet (0.03 to 0.09 cubic metre) in bulk and no more than 4 cubic feet (0.12 cubic metre). The thickness of the shell was usually 1 to 2 inches (25–50 mm). The blocks were usually made from 12 to 18 inches (300–450 mm) long, 6 to 15 inches (150–375 mm) high, and from $4\frac{1}{2}$ to 9 inches (112–225 mm) thick on the bed where they were to be bonded into brick backing.

Faience was either manufactured by the wet clay method of terracotta, described above, or it was dust pressed. This process involved the blending and sieving of a range of clays and compression of the granular dust in dry form at 12 per cent moisture content into moulds at 40 tons pressure. (In modern dust press manufacture only 5 tonnes pressure is used.) After firing, this resulted in a less durable product than that produced by the wet clay method, with a relatively porous surface, but with a more accurate final form. Plastic pressed faience tiles were usually 6 in × 6 in (150 mm × 150 mm) or 6 in × 3 in (150 mm × 75 mm) and $\frac{1}{2}$ in (13 mm) thick. Moulded edge pieces, 'trims' and borders were often 6 in × $1\frac{1}{2}$ in (150 mm × 38 mm).

Finishes to terracotta and faience

The natural surface of terracotta and faience is a *fireskin*, that is, a hard thin, vitreous, unglazed skin formed of a surface concentration of fine colloidal clay particles upon firing after the clay has been hand-smoothed or finished. Units were often finished with a *slipstain*, a thin watery paste of clay similar in texture to the underbody, brushed or sprayed on to provide a surface of a different colour or finish to the clay body beneath. *Glazing* was also used to change the clay colour and texture and in Australia and North America was even used to imitate finishes such as granite, marble and other stones. The glaze fused with the open-pored underbody upon firing as a thin, vitreous, transparent or coloured coating of glassy consistency and therefore provided a relatively impervious surface on the weather face of the final product.

Maintenance of fireskin, slipstain and glazed surfaces is critical to the well-being of terracotta and faience as the clay bodies they protect are far less durable. Their degradation can lead to one of the most complicated failure systems of all materials.

Colour of all types of terracotta and faience surfaces will vary within each surface within certain limits, dependent on the kiln type, firing temperatures, and the position and arrangement of the units with the kiln.

Proprietary terracotta and faience

Coade stone

Coade stone is a proprietary, off-white vitreous body produced from kaolinitic clay containing titanium oxide, felspar as flux, quartz as a glass-forming agent and a grog of powdered, previously fired, Coade stone.* Firing took place at 1000°C to produce a uniformly vitrified and highly durable material. Whilst the material looks, superficially, like some stones it bears no physical relationship to them.

The manufacturing principle of Coade stone was patented in the UK in the early eighteenth century by Richard Holt. In 1767 the Coade family began production of Coade stone at their pottery works on Monmouth Beach at Lyme Regis. Production was moved to a factory in Narrow Wall, Lambeth and continued until about 1840. For about sixty years the Coade factory supplied London and other areas with great quantities of architectural ornamentation such as sculptured figures, column capitals, complex and highly detailed work such as heraldry. The factory also produced an exhaustive variety of items of garden furniture, including entrance gates. Although the Coades closely guarded the secret of the materials and manufacture of their product, they were not the sole producers of this kind of material but theirs was reputedly the best. The Coade moulds were sought in 1843 and some would seem to have been bought by H N Blanchard who, after serving his apprenticeship with the firm, had opened his

*Analysis of a piece of Coade stone by the British Museum Research Laboratory indicated that the raw material was ball clay from Dorset or Devon, to which was added 5–10 per cent flint, 5–10 per cent quartz sand, at least 10 per cent grog (probably crushed stoneware) and about 10 per cent soda-lime-silica glass, which acted as a vitrifying agent.[6]

own premises at Blackfriars in 1839. Blanchard's works operated until about 1870. Between 1851 and 1875 a terracotta factory was operated by J M Blashfield, and produced Coade pieces of somewhat creamier hue.

Original models were prepared by sculptors engaged at the works from which moulds were taken for mass production. Large pieces of Coade were usually cut into manageable sections prior to firing and secured with iron or steel brackets and dowels during erection. While the ceramic body is highly durable these fixings are not; they are a major cause of Coade stone deterioration.

Doultonware

Produced, as the name states, by Doulton, this salt-glazed stoneware was produced as tiles or as modelled columns, bosses and other architectural details in a variety of colours fired once only at very high temperature (Doulton Showrooms and Studios, 1876–8, Black Prince Road). First faience components were manufactured in 1873. *Siliconware*, also produced by Doulton was unglazed and commonly of quieter colours and was used in decorative mosaics.

Vitreous fresco (Doulton)

Slabs painted in rich colours and fired to a matt surface, which were popular for large pictorial compositions in public buildings and churches.

Carraraware

By 1888 Doulton had produced a very durable matt glazed stoneware called *Carraraware*, from the white crystalline glaze which was thought to resemble Carrara marble. Carraraware was used as cladding on steel frame and concrete frame structures. Produced by coating with an enamel glaze while still unfired and then subjected to a very hot salt firing, they were very resistant to weathering. The distinctive ivory colour may be seen at the Savoy Hotel. A large range of colours was gradually developed.

Parianware

Doulton produced this dull, egg-shell finished earthenware for internal use, sometimes in intaglio and sometimes with a raised outline effect.

Polychrome stoneware

Another Doulton development which only became popular after the First World War, polychrome stoneware was a stoneware body dipped in white slip, dried and then painted in bright colours. The body was subjected to salt firing to a temperature of 1250°C, in the process of which the coloured coat fused perfectly with the body.

The faience family also includes the following types of tiles:

Encaustic tiles

These are tiles with inlaid patterns of different coloured clays. Each tile generally consists of three layers: the face, which is a slab of very pure clay of the colour required for the ground of the pattern; the body, which is a coarser clay; and the back, to prevent warping, which is formed with a thin layer of clay different from the body. The clay for the face is cut and placed on a Plaster of Paris slab which has the form of the inlaid pattern in relief, and then pressed. The coarser body and thin backing are then applied. The composite tile is then removed from the plaster pattern and laid face up. Slip clay of the different colours required according to the design is then poured into the different parts of the indented pattern on the face. After this has become hard, the superfluous clay is carefully scraped off, leaving it only in the parts originally indented, so as to form the pattern. Tiles to be glazed were dipped into a mixture of powdered glass and water after the firing, and then reheated. Inferior quality encaustic tiles had their colour applied as a transfer printed pattern.

Dry tiles
These are tiles of the same colour throughout, made by the dry process. They are glazed with a plain surface or an impressed pattern.

Tesserae
These are tiles sometimes made by the dry process and sometimes out of moist clay. They are so accurate in form that they can be laid as mosaic work in pavements.

Majolica tiles
These have raised patterns and their colour is applied in the form of an enamel or coloured opaque glaze. They have relatively low durability and were usually used for decorative wall facings internally.

6.2 DETERIORATION

General pattern of deterioration

Terracotta and faience are not properly homogeneous materials and therefore have a complex mode of material failure. Their fired clay underbodies are protected by a denser, more resilient outer surface coating of a glaze or fireskin which, if removed, may result in rapid deterioration of the partially vitrified clay underneath. Whilst this surface finish may be substantially waterproof, this may not be the case for the sides, base and top of the same piece. Where water is able to enter behind the glaze, crystallization of salts and sometimes frost action can cause failure. Historically, terracotta and faience were viewed as highly waterproof systems needing neither flashing, weepholes or drips. This supposition has proved to be untrue and water-related problems have been shown to be the most common cause of material failure.

Inspecting terracotta and faience

Particularly with terracotta the outward signs of decay do not always indicate the more serious problems within. An initial visual inspection will reveal the

A superb example of 'Arts and Crafts' design, this largely self-cleansing polychrome terracotta facade in Bristol shows many of the attractions and some of the problems of terracotta. Although largely in good condition, with colour, modelling marks and glaze surviving well, discolouration and small scale spalling is occurring around the edges of the blocks, the 'Achilles heel' of the material, where water finds its way under the glaze.

surface problems of crazing, spalling and deterioration of mortar joints, but the internal problems of deterioration of anchoring, deterioration behind the glaze and crumbling of internal webbing probably will not be identified. Each unit should be inspected externally at close range. Some information on the internal condition can be provided by an endoscope and a corrosion meter. Both require the drilling of a small hole but this can be easily filled and may avoid more destructive means of investigation.

Manufacturing faults

Several defects can arise out of the manufacturing process. Inadequate firing of the material often leads to rapid surface deterioration. This under-vitrification means that the protective, largely impermeable fireskin does not form properly, leaving the more permeable underbody open to attack from frost and soluble salts. This inadequacy is most often found in light buff-coloured units. Poor pressing in the moulding stage can produce laminations which are later highly susceptible to frost attack. Warping and cracking of the clay body may be caused by either too much water in the clay mix or by raising the kiln temperature too quickly during firing.

Glaze defects and properties

Occasionally glazes have inherent problems and become pitted or powdery as they weather. Lead glazes of the nineteenth century were fired at low temperatures and have been shown to deteriorate relatively quickly. The glazes of terracotta and faience have different physical properties from the fired clay body they coat. The glazed units leave the kiln in a totally dry state and gradually absorb moisture for the first two or three years so that the clay unit expands slightly in size. The less porous a glaze the less is its capacity to accommodate the expansion of the body. The glaze goes into tension and if its strength is exceeded, it will crack in a random hairline manner. Unless the cracks visibly extend into the porous body beneath this should not generally be considered a serious material failure. Dirt which enters these cracks cannot be removed without further damage to the glaze itself. Glaze cracking may increase the water absorption capability of the glazed unit.

The evaporation of water

Should water enter the clay body of a glazed unit from behind or through defective joints, a glaze will impede evaporation. The water is stopped at the glaze until it builds up sufficient pressure to force off sections of the glaze (glaze spalling) or portions of the glaze and clay body (material spalling). Both glaze and material spalling should be dealt with as soon as possible as they expose the porous underbody to further water entry and deterioration.

Porous joint material in a glazed terracotta or faience system is likely to be the only point of exit for trapped water.

The replacement of missing or badly damaged terracotta units should be a high-priority item as they permit water to enter the building and may also increase the structural load on the remaining pieces.

Salt crystallization damage

Water which enters unglazed terracotta and faience and which contains soluble salts will evaporate and deposit crystallized salts either on or immediately beneath the surface. Subflorescence ('cryptoflorescence') can cause the blistering, powdering and exfoliaion of the surface. Soluble salts can, of course, be introduced from many sources. Terracotta chimney stacks and terminals are notoriously vulnerable to salt crystallization damage from salts derived from flue gases, plinths and from de-icing of pavements. Similar damage may be seen where limestone dressings are placed over clay bodies so that calcium carbonate and sulphate washes into the clay.

Rusting of cramps and armature

Water which enters a terracotta system can cause rusting of the iron or steel anchoring system. Resultant damage can range from staining and spalling of the surface to cracking and loosening of the whole units which will threaten the

The hollow-block construction of terracotta units, filled with breeze concrete, shows clearly on this picture taken during the replacement of damaged Carraraware. Also in evidence is the close association with iron structural elements, and the crude nature of the backing material. The vulnerability of terracotta construction to water penetration is self-evident. Iron must be replaced with stainless steel or non-ferrous metal or, if it is to remain, be flame or abrasive cleaned and immediately treated with an epoxy resin paint or zinc rich primer.

structural integrity of the building. The majority of deterioration problems in a terracotta building are caused by rusting steel or iron. Rusting support systems are also a common deterioration problem in Coade stone items.

Early detection of a rusting anchoring system is exceedingly difficult without destructive investigation, and hence it often accompanies the beinning of repair or replacement work.

Stress-related deterioration

Where terracotta is used as a semi-structural curtain wall fixed to a large concrete or steel frame, the two may move differentially. Where there is surface loading or skin compression the terracotta blocks will exhibit compression flaking and crumble. Cracks running through many units or storeys are indication of a tension-related problem. Such cracks permit significant water entry and even if dynamic should be caulked with mastic to prevent this.

Mortar and joints

The survival of terracotta and faience depends quite significantly on sound mortar joints, whose deterioration has become a major source of breakdown of glazed terracotta and faience. The rainwater run-off over the generally impermeable surface is considerable and it is essential that the joints perform well.

Joints in terracotta and faience are typically 4.8–6.0 mm ($\frac{3}{16}$–$\frac{1}{4}$ in) wide and traditional practice has used a dense hard mortar of 1 part cement:3 parts sand. This mortar was usually stronger than the units it joined and unable to accommodate moisture and thermal movement. This meant hairline cracks often appeared between it and the clay leading to the ingress of water but not permitting its evaporation. On freezing of the water, the dense mortar resists the exerted pressures and causes spalling of the weaker clay arrises.

Inappropriate repairs

Unfortunately, damage and decay are often repaired in unskilled ways which are always disfiguring and may sometimes exacerbate the original problem. Typical of such repair methods is mortar filling with dense ready-mixed mortar coloured with stains or pigment. To avoid difficult preparation, bonding agents are added which permit feather-edging of plastic repairs and ragged patching. Sometimes paint and varnish are added to simulate glazes. Cheap repairs of this kind can extend damage by encouraging water and salt migration around the perimeter of the insertion. They are, at best, a waste of money.

6.3 CLEANING

Cleaning systems to be avoided

There are several cleaning techniques which can irreparably damage the surface of terracotta and faience, subsequently exposing the porous clay body to the damaging effects of water and accelerated deterioration. It is therefore particularly desirable to clean a trial area. All abrasive cleaning methods, the use of hydrofluoric acid, the use of other strong acids, high-pressure water and the use of metal bristle brushes can cause serious problems.

Mechanical cleaning

Air abrasive cleaning: glazes are pitted and scored by this method. On unglazed surfaces the protective fireskin may be descaled and peeled away by the normal abrasive aggregates and pressures used for sandstone cleaning. As the successful cleaning of unglazed terracotta by abrasive methods demands a very high, consistent standard of workmanship and a thorough understanding of the nature of the surface and the soiling, both wet and dry abrasive cleaning methods are best avoided.

Spinning off with carborundum discs is immensely destructive. It presents a uniform clean appearance but at the expense of the whole of the face of the block. It should never be used on terracotta or faience.

Acid-based cleaning

Hydrofluoric acid (HF) cleans terracotta and faience by attacking the complex silicates of the glaze and fireskin, thereby releasing the dirt. Therefore, the surface of these finishes is removed and etching occurs. Products of this reaction are silicofluorides and insoluble colloidal silica. Until recently the experience of cleaning glazed and unglazed terracotta and faience with hydrofluoric acid or hydrofluoric acid-based solutions has generally been bad. Many glazed surfaces have been etched and the white deposition of colloidal silica either as a bloom or in drips are clear indications of where the method has been used.

As with sandstone, extensive precautions need to be taken before, during and after an HF clean (see Volume 1, Chapter 5). Tests undertaken by RTAS during 1986 on a wide variety of glazed and unglazed terracotta surfaces have shown that there is a strong case for:

1 Using HF at a concentration of 2–4 per cent at dwell times of 5–10 minutes on both glazed and unglazed surfaces, to achieve a good but not complete clean which barely etches the glaze or removes a bare minimum of fireskin

2 Because of the tendency of terracotta and faience to soil most on the upward facing or water drenched surfaces there is a strong case for these areas to be cleaned a second time in accordance with item (1), once the general clean has been done, should this level of clean be appropriate

3 On buildings which have heavily soiled zones, tests should be conducted to ascertain what level of clean is appropriate. The amount of dirt which can be removed without damaging the surface needs to be balanced with the final appearance of the building. A strong HF clean (15 per cent) on a white carraraware surface can produce a cleaned surface where the stark contrast of the clean glaze and the remaining black soiling is visually unacceptable.

Some success in cleaning both glazed and unglazed surfaces has been had with neutral pH detergent in warm water and plastic pot scourers. This method will only remove superficial and loosely attached dirt. However, unless HF cleaners can be used with absolute confidence, methods such as these are preferred.

Alkali-based cleaning
Cleaning with caustic soda (sodium hydroxide) usually leads to the deposition of a staining white efflorescence and streaking over the terracotta or faience. The cleaning material is absorbed readily into joints and is trapped adjacent to the vulnerable underbody, where the combination of wetting and drying cycles and soluble sodium salts can readily establish a pattern of decay, especially in the form of minor spalling around joints.

Recommended approach to cleaning

Glazed surfaces
Soiling of glazed terracotta and faience may occur either in or under the glaze. Surface soiling may be very firmly bonded. Some surface soiling may be cleaned off relatively easily by water washing or using a water-rinsable neutral pH liquid soap. Plastic pot scourers are surprisingly helpful. Small areas of stubborn dirt can be moved using an emulsion of methylene chloride on a pre-wetted surface, avoiding excessive application to minimize discolouration. Soiling under or within a glaze cannot be removed without damage to the glaze.

Unglazed surfaces
Unglazed terracotta and faience surfaces are notoriously difficult to clean. It is thought this may be due to dirt forming a bond with a residue of reactive silica left on the surface during the long burn in its manufacture. Unfortunately there is no satisfactory method of removing all the dirt from heavily soiled surfaces of this kind without, in the process, removing part of the surface. The 'safe' systems are those which leave a percentage of the dirt intact. It should be remembered that the dirtiest, most drenched parts of an unglazed terracotta or faience building are likely to remain the dirtiest after cleaning. When looking at a soiled, unglazed surface it should also be remembered that unsuccessful attempts to clean in the past may have resulted in the application of coloured slurry to cover up the dirt.

Usually when cleaning unglazed terracotta and faience it is necessary to accept a partially cleaned appearance if damage is to be avoided. Poulticing with attapulgite or sepiolite clay and water achieves no improvement unless the soiling is very superficial. Poulticing with EDTA and alkaline additives in a cellulose body is marginally more successful.

The least harmful method of cleaning which produces a significant improvement in appearance is to spray with hot water before gently scrubbing the surfaces with hot water and a neutral pH soap. Scrubbing should be with compact bristle heads of the stencil brush type or a plastic pot scourer. Abrasive powders and powder-based detergents should not be used.

Unglazed terracotta and faience would greatly benefit from maintenance washes to prevent any dirt deposits building up on their surfaces.

6.4 REPAIR AND MAINTENANCE

Investigation and documentation

Before carrying out repairs to the joints or surface of terracotta and faience units, it is essential to identify the causes of deterioration, such as open joints allowing water ingress, migrating salts from a backing wall, run-off from limestone or, the most common, a corroding metal tie or armature. Remedial work can be useless if the cause of deterioration is not stopped. The eradication of metal corrosion may require major structural repairs.

During documentation of a repair programme, units to be replaced should be clearly identified by indelible marking on the faces on the blocks. Units to be repaired should not be marked with indelible materials which can be extremely difficult to clean afterwards.

Pointing

As the traditional mortar of terracotta and faience construction was as strong or even stronger than the ceramics, it is not advisable to continue its use, especially where it can be seen that it has caused deterioration. Pointing should always be carried out with a mortar which is lower in strength than the adjacent units, should permit the outward migration of water through the joints rather than through the clay units, and should be flexible enough to cope with the varied expansion of the terracotta and faience while bonding to the sides of the units and resisting rain penetration.

In addition to these conventional bedding and grouting/filling mixes a number

Pointing mixes: bedding and filling mixes

- HTI powder : lime : sand*/stone dust
  1 : 1 : 3
- Based on a coarse stuff of 1 part lime : 3 parts sand*:
  Coarse stuff : white cement
  6 : 1
- Hydraulic lime : sand*/stone dust
  1 : 3
- Masonry cement : sand*
  1 : 5–6

**Note:* The sand should be well graded down from a maximum sieve size of 1.18 mm. The mortar should be as dry as possible whilst still being workable. Lime should be to BS 890 in the form of putty. Pointing with waterproof caulking compounds such as mastics or silicone rubber should only be undertaken as a temporary crack or joint filling of proprietary materials are available which may be appropriate for the bedding of faience slabs. In particular synthetic rubber-based materials have been extensively used. Two-part, water-based ceramic tile adhesives are available which may be applied to a wall surface in thicknesses up to 12 mm ($\frac{1}{2}$ in). Substrates must be dry and clean and slabs must be fixed in position before surface drying occurs. Grouting of tiles bedded in synthetic mortars may be carried out using conventional lime-based grouts or, in some cases, a coloured or white silicone rubber may be appropriate.

To match untouched joints, clean a small sample of dirty joint face with 10 per cent hydrochloric acid making sure the joint is wetted first and thoroughly washed off afterwards. Where re-pointing has already taken place a sample will need to be obtained from behind to ascertain the intended colour of the joints. Traditionally the joints in terracotta and faience as in ordinary ashlar work should be finished flush or slightly concaved but not struck.

Old mortar can be carefully removed using a mason's saw or tooth-edged chisel but a hacksaw blade will suffice. Diamond-tipped power saws and carborundum wheels are extremely difficult to control and should only be used if the unit edges are properly protected. Terracotta and faience are extremely weak in tension and great efforts must be made to reduce the risks of levering against the arrises when cutting away old mortar. This is also the case when chasing or cutting out areas of friable material.

The procedure of pointing is described in detail in Chapter 4, this volume, 'Pointing stone and brick'.

Joints should be pointed using purpose-made pointing irons capable of fitting into the joint to ensure good compaction without spreading the mortar on the face of the work. Any mortar droppings must be removed at once and not allowed to harden on the surface.

Preventing ingress of water

All items in terracotta and faience construction which have the function of preventing ingress of water must be carefully maintained. Because of the brittle nature of existing dense mortar between terracotta and faience units, it may be advisable to protect horizontal surfaces on cornices, parapets and elsewhere with lead flashings. Where there is evidence of cracking and the rusting of armatures, it may be appropriate to provide weepholes in the underside faces of overhanging terracotta blocks for the purposes of draining before grouting or pointing.

Patching using plastic repair

Damages to small areas of terracotta and faience which are restricted to individual blocks can receive plastic repair in an appropriate mortar. This technique is suitable where the glaze or fireskin and only part of the underlying fired clay body has fallen away.

It is always advisable to carry out a test repair in an inconspicuous place.

To repair a minor chip or spall without undertaking any preparatory cutting must involve the use of an adhesive admixture. Small repairs can be carried out using cellulose nitrate as the adhesive, which will not discolour. Polyester pastes or epoxy putties are also used on a small scale. Rather larger areas are often filled with feather-edged mortar and such repairs will always be vulnerable to early deterioration and are never visually satisfying. Such spalls should therefore be left alone unless their size and position makes that unacceptable, in which case a neat square repair should be formed.

A line should first be cut with a glass cutter through the fireskin or glaze at right angles to the edges of the block. Once this cut is made a masonry drill can be used to excavate the area of damage. With care, almost all the material can be removed in this way but the corners may need to be cut with a small hacksaw blade. If possible at least two of the edges should be slightly undercut to provide a good key for the repair. The depth of repair will, of course, vary in proportion to its surface area but in general not less than 10 mm is acceptable.

Mortar samples for unglazed surfaces should be prepared to match the colour of the weathered terracotta or faience. The sample must be thoroughly dry before a decision is made. The colours should be obtained through the use of a soft staining sand or limestone dust and/or brick dust. At least 50 per cent of the aggregate should be a sharp sand although the maximum size of aggregate cannot normally exceed 600 microns without giving the appearance of a gritty and therefore inappropriate matrix. The binder can be 1 part masonry cement to 5–6 parts aggregates. The use of masonry cement permits a higher percentage of aggregate to be used and reduces the neutralizing colour effect of other binders. Alternatively if a pale, pasty appearance is required a mix of 1 part HTI:1 part lime:3 parts stone dust may be better.

To avoid fine shrinkage cracks in and around the repair it is essential to achieve a clean, dust-free and damp surface by hand spraying before the mortar is placed. The mortar should be as dry as possible whilst still being workable. Deep cavities should be packed with broken tiles and mortar before beginning repairs. The depth of repair should not exceed 10 mm in one application and spraying must precede each application of mortar. Unlike other forms of plastic repair it is useful to finish the repair flush using a small steel trowel. The laitence so produced can, with practice, simulate the fireskin of unglazed terracotta or faience quite well. On unglazed work the repair and surface in which it is placed may be finished with one or two thin applications of microcrystalline wax.

The repair of glazed surfaces should be carried out in a similar manner although it will not always be important to achieve the glaze colour with the colour of the repair mortar. This can be done with a tinted UV-stable polyurethane glaze.

The standard of repairs

Several proprietary compounds exist for the repair of terracotta and faience, based on cement and adhesive additives. Experience with these in the UK has not been sufficiently long for positive recommendations to be made. Many of these compounds achieve colour matching by the use of pigments and these are expected to have the usual colour-leaching problems. Many repairs conducted so far have not been preceded by proper cutting out and preparation, resulting in unsightly and unsound feather edging. Good workmanship is an important part of good terracotta and faience repair.

Mimic glazes

Mimic glazes are rarely a total success and for this reason special consideration should be given to replacement of all but the most superficially damaged blocks. The glazes on terracotta and faience are vapour-permeable to a degree and synthetic glazes are not generally successful in matching the original glaze in this respect.

Experience to date in this field is limited but materials applied to recreate the glaze have been gloss polyurethane 'varnish', acrylic paint and clear epoxide coatings. All three systems are susceptible to degradation in ultraviolet light and may in time begin to powder, 'chalk' and discolour. The materials are usually of low viscosity and will run on vertical surfaces, so care must be taken to protect underlying material from drips and stains. Ultra violet-resistant clear polyurethanes are now available and beginning to be used in glazed surface repair. Certain suppliers can produce a matt polyurethane on request. The long-term performance of these materials is not yet proven and will be of great interest. Two-part epoxy marble fillers have been used successfully to repair small defects in the white and off-white glaze of carraraware. Acrylic resins made up as a 10 per cent solution in acetone and industrial methylated spirit (1:1) may be used to consolidate deteriorating glazes in localized areas.

Pinning loose pieces in situ

Anchoring systems using patented mechanical fixings with or without adhesives may be used to secure unstable units. Various systems are described in *BRE Digest 257* – 1982 'Installation of wall ties in existing construction'.

Where small, exposed, solid decorative features or the faces of units have broken away from the facade or where the cement backing to faience has failed, it is possible to use these fixings without the necessity of dismantling the wall. Holes between 6 mm and 15 mm in diameter are drilled through from the face of the loose item into the resilient backing wall or substrate and the ties are then placed and either fixed mechanically or grouted in with thermoset resin. The countersunk holes in the finished suface of the units can be made good subsequently with a plastic repair.

Sealing cracks

In rare cases where cracks are not associated with continuing structural movement or metal corrosion, for example, physical impact or frost action, they may be gravity-grouted with a colour-matched thermoset resin of low viscosity. Active cracks which need to be temporarily sealed to prevent the ingress of water should be filled with a polysulphide caulk. These waterproof compounds are not appropriate re-pointing materials and should not be used as such.

Bonding and stitching of broken units

Where areas of terracotta or faience have to be dismantled, units or features may crack or break during the course of the works. The broken material should then be removed from the wall and subsequently drilled and dowelled. Adhesives should be applied to clean broken edges and these should be epoxy resin.

Where necessary small-diameter pins should be inserted as dowels or location registers before the broken pieces are brought together and bonded with the thermoset resin. All fixings used for surface repairs should be of stainless steel, non-ferrous metal or polyester.

If the void at the back of the unit is filled with a dense concrete this cannot wholly be removed without damaging the clay body; the rough edges should be dressed off flush with a claw chisel or power saw to keep the arrises and edges clean to facilitate rebonding in the wall. If this concrete packing provided a mechanical lock between adjacent blocks, ties should be used for refixing.

Bonding and stitching of broken pieces should be carried out only by experienced craftsmen or conservators.

Replacement with new units

Terracotta and faience units which have spalled severely, thereby losing much of their material and structural integrity in a wall, should be replaced. Whole blocks or tiles should be replaced and not parts of units. A number of manufacturers are able to produce new terracotta and faience elements.

Terracotta and faience replacement companies have a limited range of clays available to them. For unglazed terracotta it is not always possible for one company to achieve the required colour from the clay body available to it and a glaze may need to be used. It can be worthwhile enquiring to find another company which is using a suitably coloured clay. It must be recognized that, should it be necessary to replace an unglazed block with a glazed block, any damage to the glaze will reveal the clay body beneath. The skill of matching glazes is, however, well developed.

The process of producing replacement terracotta and faience is delicate and time-consuming and work should be programmed with this in mind. Working with terracotta requires masonry construction skills; anyone experienced in the restoration and fixing of stone should, therefore, be able to undertake the cutting out of defective pieces and the fitting of replacement. There are a few masonry companies that specialize in terracotta fixing.

Terracotta and faience should be replaced with like materials. Alternatives such as precast concrete, artificial stone, GRC and GRP are usually cheaper but should only be considered for emergency replacement where more ceramic material would be at risk if intervention did not take place and where the use of a substitute does not become the excuse for not using the correct material. Each of these substitutes has notable disadvantages and all weather in distinctly different ways from terracotta.

Because of the method of manufacture and firing variations will also occur, but the manufacturer should ensure that the units provided are thoroughly burned and free from cracks, air bubbles and other defects; that arrises are sharp and that the enriched work is clean and well undercut. The units should be true and even on each face and of even thickness throughout. It should be remembered that the effects of age and weathering cannot always be reproduced and that replacements will be obvious for many years. Attempts at artificial weathering or soiling should nevertheless be vigorously opposed.

The problems of cleaning terracotta without damage have lead to the decision to retain the sound, but heavily soiled, units shown in this illustration. Replacement blocks have been carefully matched, allowing for firing shrinkage, and have been bedded in a rather weaker, more elastic mortar than the original. Although this may be considered very unsatisfactory from a visual point of view, this solution is 'correct' in conservation terms.

In replacing units, it is advisable to inspect the anchors and reinforcing armature thoroughly for any signs of corrosion. All rust should be removed and the metal cleaned and coated with rust inhibitor and a water tolerant paint. The corroded fixing may need to be replaced by non-ferrous stainless steel or polyester components which provide an anchoring sysem similar to the original.

New units should be wetted thoroughly before placing to control suction of the mortar. Voids between units and their armature should be filled completely with one of the mortars described on page 79, 'Pointing', with the addition of 1–2 parts clean, broken brick aggregate. A suitable final mix would be 1 part binder: 7–8 parts aggregate. The block filling should be placed without tamping or ramming. Where mechanical bonding or void filling with concrete is unnecessary, polyurethane foam may be used to fill the back of the clay boxes in order to prevent moisture penetration inside the unit. Joints should match the width and profile of the original and be filled with a mortar also as explained on page 79, 'Pointing'.

Faience slabs should be applied to their backing wall in accordance with the procedure for tile cladding described in section four of BS 5385: Part 2: 1978 'Code of Practice for Wall Tiling: External Ceramic Wall Tiling and Mosaics'.

Water repellents

Only exceptionally would the use of silicone or other water repellents be recommended. However, where the joints are basically sound but have fine shrinkage cracks which are permitting the ingress of water, a 'wet' treatment with a silicone water repellent to BS 3826 (Class A) or BS 6477: 1984 (Class 1) may be used. (See Volume 1, Chapter 10, 'Colourless water-repellent treatments'.) 'Wet treatment' entails wetting the facade with a hose or water lance and applying a flood application of silicone to the wet face. Because so little silicone is taken up by the wet surface of the blocks it is concentrated in the hairline gaps at the joints.

Consolidants

Acrylic resins in acetone-methylated spirit, acrylic resins in alkoxy silane and catalysed alkoxysilane have been used on a limited scale and largely experimentally. Experience to date does not suggest that they are likely to be used extensively or in situations other than local areas of decay.

REFERENCES

1 Atterbury, Paul and Irvine, Louise, *The Doulton Story*, Exhibition Catalogue/Booklet, 1979, Royal Doulton Tableware Ltd, London Road, Stoke-on-Trent, England.
2 Building Research Establishment, *BRE Digest No 257*, 'Installation of Wall Ties in Existing Construction', HMSO, Garston.
3 British Standards Institution, *BS 5385: Part 1: 1982 Code of Practice for Wall Tiling Part 2: External Ceramic Wall Tiling and Mosaics*, Appendix G, 'Cleaning and Surface Repair of Terracotta and Faience'.

4 BS 6270, *Code of Practice for Cleaning and Surface Repair of Masonry*.
5 Fidler, J A, *The Conservation of Architectural Terracotta and Faience*, 'ASCHB Transactions', Vol 6, 1981.
6 Freestone, J C, Bimson, M and Tite, M S, *The Constitution of Coade Stone*, 'Ceramics and Civilisation, Vol I: Ancient Techniques to Modern Science, W D Kingery (ed).
7 Kelly, Alison, *Coade Stone in Georgian Architecture*, Vol 22, pp 71–101.
8 *Preservation Briefs No 7, The Preservation of Historic Glazed Architectural Terracotta*, US Department of the Interior, Heritage Conservation and Recreation Service, Technical Preservation Services Division.
9 Snell, Peter, *The Conservation of Architectural Terracotta*, Thesis for the Architectural Association Postgraduate Diploma in Building Conservation, 1983.

See also the Technical Bibliography, Volume 5.

7 THE REPAIR AND MAINTENANCE OF COB, CHALK MUD, PISE AND CLAY LUMP

7.1 EARTH WALLING

This chapter considers the use of unbaked clays and soils in the construction of walls.

Earth walling is based on the principle that when certain suitable soil materials with the right moisture content are tightly compressed, they cohere to form a fairly hard, strong, and solid body. The necessary cohesion can be obtained by external compaction of a relatively dry earth mix in shuttering (pise), or by the natural drying out of water from a wetter mix either placed directly in a wall (cob or chalk mud) or which has been formed into blocks which are then built into a wall (clay lump). In each case, once the wall is constructed, the cohesion lasts only so long as the materials are kept dry. Earth walling is, therefore, a high maintenance form of construction. The continuing operation of wide overhanging eaves, a continuous surface coating, an isolating plinth and effective rainwater disposal are critical to its survival.

History and distribution of the types of earth walling

The local craft of earth building is one of the oldest forms of construction in Britain. The various forms of clay walling are found in areas where the soil is comprised of clay, sand or chalk or a combination of these. The type of soil available was one of the important determinants of the form of construction used.

Cob buildings are found from Cornwall to Hampshire, the greatest number being in Devon. Towards the west, the colour of the material is that of the red-brown clayey subsoil. In Wiltshire, West Hampshire and Berkshire the walls are creamy white as the local chalk is the basic ingredient. Here the terms 'cob' or 'chalk mud' are used. In South Buckinghamshire 'wichert', a local clay with a high chalk content was used to produce walls of notable inherent stability. Other areas of cob construction are found: in the East Midlands (dark yellow earth); the Solway Plain of Cumbria ('clay daubins': small farm houses made of shallow courses of clay separated by layers of straw); Wales (cottages and farm buildings made of yellow clay and/or grey mud containing flakes of stone); and the eastern counties of Scotland ('clay daub'). There are references to buildings of mud or clay in many parts of Britain, in all historical periods. The Poor Law Amendment Act of 1834 stimulated the most recent revival of cob construction for cottages.

Clay lump buildings are very common in central and southern Norfolk, present but less common in south Cambridgeshire and west Suffolk. Many farm buildings in these areas are tarred while the cottages are lime plastered. Coal tar began to be used in the 1850s when it became a cheap by-product of the gas works. Clay lump cottages can also be found in north-west Essex, Bedfordshire and eastern Hertfordshire. There is little evidence of standing buildings of clay lump earlier than the eighteenth century, and most are of the early nineteenth century.

The *pise* process was traditional in the Rhone Valley of France and began to be used in England after 1790. However, pise occupies a minor place in the history of earth building in Britain. It is not possible to indicate its regional distribution other than to say it was mostly used in southern England at a higher social level than the traditional cob walling.

In Norfolk, Suffolk, Cambridgeshire and Essex the 'puddled clay' method of construction can be found. Clay-based earth was placed in a soft state between shutters in lifts of 18 in (450 mm) high. Walls of about 13 in (325 mm) were produced, with inner and outer surfaces being very true. Timber frame construction was used above eaves height of these buildings. Mixtures of the various earth wall systems can also be found, for example, cob and puddled clay, stone and puddled clay.

The use of the earth-building processes declined from the 1850s mainly because bricks were becoming cheaper (improved productivity and abolition of the brick tax).

Cob and chalk mud

Cob walls and chalk mud walls are both erected by the same unshuttered method of construction. The term '*cob*' usually describes walls comprised of soils containing clay and sand with straw added during the mixing to assist in the drying out and to distribute shrinkage cracks throughout the wall. The term '*chalk mud*' usually refers to walls of crushed chalk, that is, their composition is crushed chalk only, chalk and straw or chalk, clay and straw. The walls are built in the absence of shuttering by the simple process of pitching on a soft but cohesive mixture of the materials mentioned, in layers, the wall surfaces being pared down with a flat-backed spade to form a fair face as the work proceeds.

The composition of the soils depended on what was immediately available. Cob walls usually contained a well-graded mix of clay (about 20 per cent), silt or fine sand, coarse sand and gravel with some straw. Proper grading was also necessary for chalk mud walls where the aim was a conglomerate of small chalk knobs (usually not larger than 40 mm) and very small lumps cemented together by a matrix of plastic chalk, the whole forming as dense a mass as possible. As cob is dependent on fine clay particles for strength, so chalk mud depends on a proportion of fine chalk dust. The chalk mud walling of Wiltshire consists of well-trodden chalk lumps and straw. In Hampshire some walls were made of 3 parts chalk to 1 part clay while others of the last century had slaked lime added. Lime was also added to some cob walls.

The chalk and clay were usually dug from the subsoil in the autumn and left in the winter frost to achieve the necessary grading for a strong and stable wall. The material was mixed on the ground next to the wall, trodden by labourers and sometimes horses, the straw being added during this process by scattering in a series of thin layers, each one being placed before the mixture was turned. In a Hampshire chalk mud of 1 part clay to 3 parts chalk, one hundredweight (45 lb) of straw was mixed to every ten perches. (One perch is $16\frac{1}{2}$ ft (4.8 m) long by 1 ft (300 mm) high.) Less straw was used in pure chalk walling than in a chalk and clay wall and was sometimes omitted altogether. Only enough water was added to make a plastic mix which 'stuck like glue to the workmen's boots' (reference 14).

A well maintained chalk mud wall, 450 mm (18 in) thick on a brick and flint pinning with a wired straw thatch capping on a sawn oak frame. Although substantially weathered off, this wall was protected on both sides by a buff-coloured lime and tallow wash. This wall can survive indefinitely provided the thatch is kept in good order (probably every 15 years) and the limewash is renewed (probably every 5 years) and the soil level is kept well below the top of the pinning.

Up to two hours was often required for proper mixing and it was often reported that cob and chalk mud mixes 'can hardly be too much trodden' (reference 14).

Cob and chalk mud walls consolidated by virtue of their own weight and by drying out. Their strength depended on the proportion of materials, thorough mixing and the compaction which could be achieved by treading. Only in rare cases during the nineteenth century where some lime was added was extra strength provided by a weak setting of the mix.

A pinning (plinth) 14–24 in (350–600 mm) wide and 1 ft (300 mm) out of the ground was built of flint, stone, brick, rubble, boulders or pebbles. Its purpose was to prevent damage from rain-splash and to isolate the cob or chalk mud from wet ground. The wall was built up on the pinning by a man pitching the mixed stuff to his partner on the wall, who caught the lumps on a three-pronged fork, placed them, and trod them down. At this stage the material projected $1\frac{1}{2}$ in (38 mm) beyond the face of the footing wall. The wall was built in courses 1–2 ft (300–600 mm) high. As the material had to be hard enough to walk on before the next course could be placed, up to a week was often required for a course to dry. In cob walls a layer of straw was sometimes laid over each course to assist the drying process. Lintels were bedded in solid on cross pieces and the openings cut later. At intervals the waller pared down the surfaces with a flat-backed spade.

Once completed the wall was left to dry out thoroughly. It was recommended that walls were built from March to September, that the internal fitting and plastering was carried out during the winter and that the external rendering was applied no less than one year (preferably two) after completion to allow the walls to become reasonably dry.

Chalk mud construction was used extensively for boundary walls. (At the end of the Napoleonic Wars a profusion of chalk and boundary walls was built in Hampshire to protect the cottage gardens from being raided by troops disembarked at Southampton.) These were usually capped with thatching with a wide overhang. The external wall surfaces received the protection of a lime render or several coasts of limewash (see page 00). External surfaces which did not receive this kind of protection are rare.

Pise (rammed earth)

Pise de terre, or rammed earth, is a semi-dry method of building walls. It consists of ramming earth of low moisture content between boards of climbing shuttering. (Stabilized earth is a post-World War Two version of pise and includes the addition of cement, bituminous emulsion or other agents for walls which require greater strength and weather-resistant properties.)

A favourable earth mix was thought to be 25 to 30 per cent clay to 70 to 75 per cent sand and gravel, in a well-graded mix with the minimum of air voids. Chalk was used in Hampshire; over half of the remaining chalk boundary walls in Kings Somborne were built by the pise method, superseding the chalk mud method during the period 1812–1841 (reference 11). Pure chalk was pounded and graded from 32 mm down into a well-graded mix, often including a few flint chippings. No straw binder was included in the mix nor left at the horizontal joints of walls of either soil category.

The moisture content of the soils used was critical to the stability of the pise wall. If it was too dry it would not consolidate when rammed and if it was too wet it pugged, stuck to the rammer and shrank upon drying, weakening the structure. The desired moisture content for optimum ramming conditions was found to be from 10 per cent by dry weight for a sandy soil to approximately 20 per cent for a clay type of soil and for chalk. A rough guide for checking this was to take a handful of the earth and squeeze it in the hand. If it just adhered together it was deemed of the right consistency. A plastic state indicated an excess of water and crumbling meant too little water. A second method was to roll out the earth on a flat surface with the palm of the hand until it was about $\frac{1}{8}$ in (3 mm) in thickness. If at this point it broke in two, the moisture content of the test soil was about right. If it crumbled earlier it was likely to be too dry.

A plinth was built, as for cob/chalk mud, but with square trenches across the top at 3 ft (900 mm) intervals to receive the putlogs or transverse timbers of the frame which was then erected. The clay soil or the chalk was dug and sieved. Wet chalk was not dug and no extra water was added. Only sufficient chalk was dug for one day's work. Soil blending occurred where a mix of 70 per cent chalky gravel to 30 per cent fine loam was used. The thickness of pise walls varied from 250–450 mm (10 in–18 in) according to the load to be imposed on them.

The soil was shovelled into the shutters in layers 75–100 mm deep, rammed using blocks of elm with alternate blows by two workers, starting near the shuttering and working towards the centre of the wall. The soil was rammed until less than half the original thickness, when ramming marks no longer showed. The process was repeated until the shutter was full. The shuttering was dismantled and moved further along, over-lapping the end of the previous section which was left at an angle of 60 degrees. Because no time was required for drying out, successive courses could be carried out immediately. When completed, a building was roofed and left to dry for two to three months. Finally the putlog holes were filled and the walls rendered or limewashed.

An interesting version of pise wall construction occurred in Winchester in 1841–3 where a series of town houses were built with 16 in (400 mm) thick solid walls of rammed chalk with $4\frac{1}{2}$ in (225 mm) brickwork as permanent, external shuttering (reference 11).

Clay lump (earth blocks)

Clay lump describes the construction of walls of large unburnt blocks of clay-type soil and is found principally in East Anglia. The blocks were formed by compacting the clay into wood moulds. Although larger in size than bricks, the initial preparation, moulding and drying processes followed traditional brick-making processes.

The soil mix for clay lump blocks generally contained at least 50 per cent of sand although 85 per cent sand/gravel is also to be found, ie 5–15 per cent clay content. Soils containing a high proportion of clay also included binding materials such as straw or grass. The clay was dug and left exposed to the winter frosts. On the ground it was mixed with straw and water, trodden by labourers or horses and turned with farm implements until it became a sticky mass which

New clay lump construction showing 18 in (450 mm) × 9 in (225 mm) × 7 in (175 mm) blocks on a brick plinth. The blocks are bedded in lime and clay, plastered in straw reinforced lime sand plaster and finished with a lime tallow wash. Alternative traditional finishes include lime and oil or sanded tar.

could be picked up with a fork. Only enough water was added to make the soil conveniently plastic, enabling it to be pressed into a mould with the feet and a flat tamper and to hold its own shape once the mould was removed. A moisture content of 16–20 per cent would be typical. The timber moulds, which had neither top nor bottom, were wetted and laid on a thin layer of straw on the ground. Once properly filled, the excess earth was cut off flush with the top of the mould with a wire bow as in brick-making and the mould was removed to form the next block. After turning out, the blocks were left to dry for two or three days, carefully protected from the rain. They were then turned up on edge for further drying of a week or so and finally were stacked at 45° with minimum contact so that the air could circulate freely round the greater part of all the sides.

Clay lump blocks varied in size from 18 in (450 mm) long, 9–12 in (225–300 mm) high and 4–7 in (100–175 mm) deep.

A rare example of lump blocks made from chalk occurred in 1932 at Quarley, East Anglia, where Albery (now Etekwen), Thimble Hall and some nearby cottages were constructed of these clay lump walls built on to a base wall of brick or stone about 18–24 in (450–600 mm) high (reference 14).

Construction usually took place during spring and summer. The blocks were laid in the same way as ordinary bricks, the mortar joints being just thick enough to level the course. Mortars used were similar to the earth used for the blocks, but at times contained some lime. The walls produced were thinner than those of cob, chalk mud and pise. Clay lump walling is a masonry construction.

Farmhouse and agricultural buildings in East Anglia were often treated externally with coloured limewash, tarred only, or tarred, sanded and then colour washed.

Protective coverings for the surfaces of earth walls

The external wall surfaces of cob, chalk mud and pise were almost invariably protected by lime renders and/or limewashes. One or two coat lime renders of 3 parts sand to 1 part lime putty were common. The surface was plain, marked out to resemble ashlar, or roughcast. Renders of crushed chalk were used in chalk areas and sprayed with limewater (reference 8). Render and unrendered surfaces were often coated with limewashes including additives such as tallow, alum, powdered glue and skimmed milk, and tinted with pigments and occasionally copperas (ferrous sulphate). Salt was at times added to renders and washes to retard the drying out process and assist the slow carbonation of the lime.

The components and combinations of the protective coatings extended to lime renders containing linseed oil and horsehair coated with a mixture of 10 lb (4 kg) tallow added during slaking to a bucket containing one-third quicklime and two-thirds water, applied hot. Mud plasters of 3 parts sand to 1 part clay were also used (dagga or dogga plasters).

From the mid-nineteenth century onwards a mechanical key was sometimes provided for rendering by wire mesh secured with wire into the wall or by a random pattern of long nails driven into the wall.

Several early twentieth-century references on earth walling recommend the addition of Portland cement to improve the waterproofing properties of renders.

This was not traditional practice and is not recommended for repairs. A traditional method of waterproofing was to apply hot coal tar blended with sand. To avoid the risk of bleeding, limewashes were not applied for several months after a wall had been tarred, and then only on a well-sanded surface.

7.2 EVALUATION OF SOILS

Rebuilding, infilling and repairing earth walls requires some knowledge of the type of constituents which are suitable and their performance characteristics. Simple analysis of surviving fabric is also desirable.

Constituents of soils

Soil constituents are normally classified by size only.

Constituent and particle size		*Method of identification*
Clay:	below 2 microns	Sedimentation analysis
Silt:	2 microns-20 microns	Sedimentation analysis
Sand:	20 microns-2 millimetres	Sieve analysis*
Gravel:	above 2 millimetres	Sieve analysis*

*Coarse sand, 1.18 mm sieve. Medium sand, 600 micron sieve. Fine sand, 300 micron sieve and above 150 micron sieve. (Based on BS 1377: 1975)

Due to the specific characteristics of each solid particle the proportion in which they are found in a soil will influence the behaviour of that soil should it be used in earth walling. To *clay* is attributed the property of cohesion or binder in the mix. *Silt* is an inert filler which also has some binding properties but to a lesser extent than clay. *Sand* functions as stabilizer of the clay and silt and improves the behaviour of dimensionally-unstable clays. Some soils naturally contain the required proportion of fillers to binders while others need to be blended.

The traditional practice of dealing with the problem of drying out shrinkage in cob and clay lump walls was to keep the clay content at the lowest possible level by adding extra aggregate, to keep the water content to the practical minimum, and to add fibres such as straw. Not all clay soils required all three precautions – some cob walls in Devon are fibre-free because a stable (low-dimensional change) clay was used. The very initial low moisture content of pise soil reduced shrinkage contraction to a point where it ceased to be a problem.

The performance of some cob wall soils was improved by the addition of cow dung which acted initially as a plasticizer and subsequently as a binder inhibiting the dispersion of the clay in water. This adhesive role improved the weather resistance of the cob.

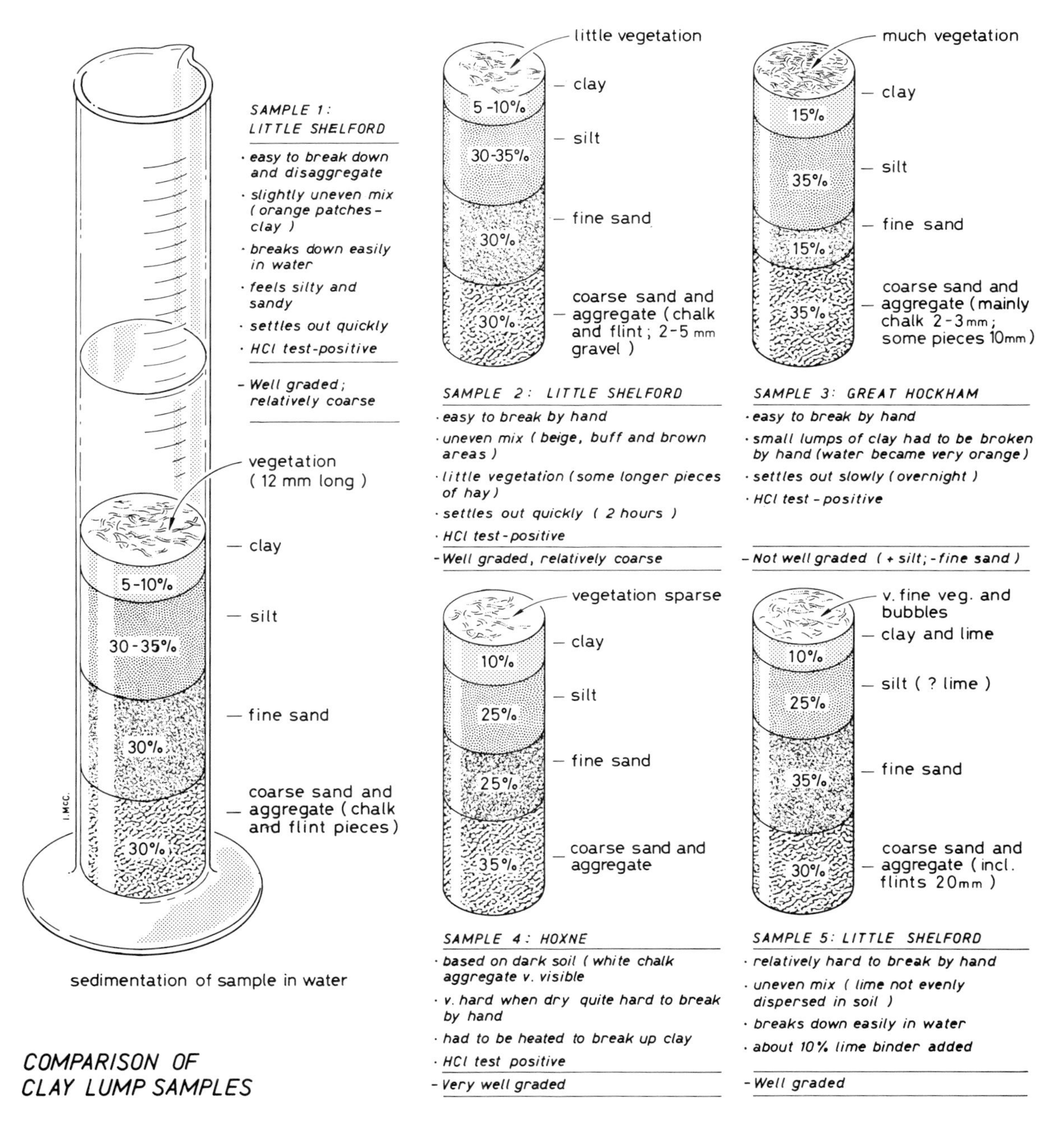

Figure 7.1 Sedimentation analysis of clay lump samples

Dung

The significant constituent of dung, especially cow's dung is a mucus which reacts with lime to form a gel. The gel has the desirable effect of supporting the lime and sand until strength due to carbonation is achieved. The gel also stabilizes the clay mineral wafers. Where a surface coating which sharply reduces the passage of moisture vapour is applied, condensation takes place behind it and the mucus gel swells. The forces involved in the swelling break down the internal strength of the material. Such surface coatings also prevent the contact of atmospheric oxygen and promote the activity of anaerobic bacteria which degrade the mucus. Such coatings should, therefore, never be applied.

Chalk

Chalk is a pure limestone (up to 98 per cent calcium carbonate) white to pale grey in colour and very fine grained, uncemented and soft where it occurs in southern England.

Weathered clay layers for walling were dug up with a pick. The lumps were left exposed during the winter to be 'frosted'. The initial breakdown of the lumps was followed by a long treading and mixing operation which broke down the wet and fragile chalk into small pieces. The quality of the wall was directly related to this thorough and patient compaction.

Analysing chalk in walls

Chalk mixes were generally coarser than the clay–soil mixes used in earth walling. Their particle size distribution is best determined by sieve analysis.

Tests for clay soils

Replacement soils should repeat the proportions of clay, silt and sand in the case of clay soils and the particle and lump size distribution in chalk walls. Laboratory procedures for this are not always available or necessary. Several field methods exist which are not highly precise but which provide adequate information. An unweathered sample should be selected for analysis.

Visual and tactile analysis

Visual inspection of an earth wall sample will reveal the type of soil, most of its components, and the final result of the mixing. Particularly with soils containing clay a lot can be learned by feeling the soil: how easily it crumbles and resists surface damages; whether a dampened sample can be rolled into a ball and what happens when this is squeezed; whether the dampened sample can be rolled into a 50 mm thick sausage and bent; whether an even damper sample feels smooth and pasty or rough and gritty; whether it takes a polish; and whether a very wet sample sticks the fingers together. The speed with which a sample absorbs a drop of water can provide further information about the clay content. This kind of analysis can provide a qualitative comparison between original and replacement soils.

Particle size distribution
For an approximate estimate of the particle size distribution of a soil, take a representative sample of dry soil, dissolve this in excess water in a tall volumetric flask, shake this mixture thoroughly and let it settle. The soil components will deposit in layers in order of density. Any organic matter will float on top and after five minutes the gravel and sand particles will have fallen to the bottom. After eight hours only the clay remains in suspension and the silt will have settled on top of the sand. After forty-eight hours the clay should also be deposited. Then the height of each layer can be measured off the side of the flask and the proportion of each particle size component estimated. If any excess clear water is syphoned off and the sample dried, a more accurate estimate of the clay proportion can be determined as the clay will form a sticky layer on top with an easily definable thickness. A thoroughly dried clay layer will cup and fracture.

Binder: aggregate proportions
A simple analysis of the soil in the ground or in a wall to show how much sand and clay it contains may be made by the following method. Take about 3 lb (1.5 kg) of soil from which all the large stones have been removed. Heat it to drive off all moisture and weigh it. Put it into a large, shallow pan or a bucket, cover it with water and gently stir. When the water becomes cloudy with the clay and silt, pour it away and renew. Repeat the washing step until the water no longer becomes cloudy, showing that there is no more clay and silt in the mixture. Then dry the sample and re-weigh it. The result will show the percentage of sand in the soil. This test can be carried out more easily by volume and the results will be roughly equivalent.

Binder: aggregate proportions – performance test
Make a roll of soil dampened to plastic consistency which is 25–30 cm (10–12 in) long and 18 mm ($\frac{3}{4}$ in) in diameter. Place this in one hand and gradually ease it over the edge of the hand. Take note of where the overhang breaks off as this will indicate the adhesion and hence amount of clay in the soil. This test is useful when comparing an unweathered sample from an existing wall and a sample of a replacement soil. Prepare a damp roll of each. Once the two rolls break at a similar point a replacement soil of similar composition has been found. If the replacement roll breaks too soon it has too much sand and needs more clay; if it breaks too late it has too much clay and requires additional sand.

7.3 THE DETERIORATION OF CLAY AND CHALK WALLS

Wall bases and wall heads
Water is the principal enemy of all earth walling, especially those incorporating a high percentage of clay. Excess water completely disperses clay binder and the remaining materials are easily washed away. Chalk walls also lose cohesion and crumble in the presence of excess water. Most frequently, damage occurs through

rain penetration at wall heads, inadequate drainage at the wall base and rain splash, especially where the ground level has risen. Capillary rise is low in earth wall materials and generally does not reach above 300–400 mm.

A high moisture content can greatly decrease the tensile and compressive strength of clay-based earth walling. The bases of the walls, in particular, where the maximum load is concentrated, tend to collapse once they become damp or saturated with water. Plinths of brick and stone are intended to provide protection from rain splash and to avoid direct contact of the earth walling material with the wet ground; it is essential to maintain the full original height of the plinth above ground level and not to allow build-ups of soil against it.

Wide eaves overhangs were designed to throw rainwater well clear of the wall surface and this detail should not be modified. Tile, slate and thatch copings should be regularly maintained. When a tile is lost, for instance, it should be replaced immediately or a temporary cover placed over the top.

Surface treatments

Where the original surface of the earth wall was limewashed or lime-rendered and washed these finishes must also be renewed when necessary to avoid deterioration. Renewal must match the original material. Cement or cement:lime compo mix renders must not be used and even modern vapour-permeable paints are not admissible. Any of these regrettably common practices can positively exacerbate the rate of deterioration by trapping moisture and encouraging the breakdown of the surface. (See Volume 3, Chapter 7, 'Limewashes and lime paints' and Volume 3, Chapter 4, 'External renders'.)

Bees, rats and plants

Masonry bees are attracted to soft, thick walls and large numbers can contribute to the weakening of a wall. Even more destructive and undesirable are extensive rat runs which are sometimes found in earth walls, especially in farm buildings. The addition of crushed glass and crushed shale were traditional rodent preventatives. Current infestations of either kind may require the assistance of the local pest control authority. Voids can subsequently be filled by tamping in a weak hydraulic lime mix as detailed below.

Plant growth on earth walls is unwelcome if excessive, but many garden walls were built as, or have become the support for, decorative or fruit-bearing creepers or vines. If such walls need repair it is sometimes possible to lift the bulk of the growth away from the wall with minimum cutting. Reinstatement should include the provision of a light climbing frame of treated wood and non-ferrous or galvanized fine wire mesh fixed between plinth and eaves and not nailed into the earth wall.

Leaning walls

In general, it must be appreciated that earth walls have low tensile strength and will tear readily due to stress caused by local settlement at ground level or around openings, especially when the wall earth has been weakened by saturation. Leaning earth walls can sometimes be restrained with tie rods, but successful traditional ties usually have oak face plates of at least 15–18 in (375–450 mm) width and 3–4 ft (900–1200 mm) height. Standard masonry plates are likely to tear through the wall if tie rods are applying restraint unless a horizontal stitch has been introduced. The construction of buttresses has also been attempted at times to oppose the lean of an earth wall, but unless these are placed exactly where they are required and their construction is of high standard and does not encourage water retention, this is to be discouraged. The additional load of a brick buttress on a soil substrate weakened by years of water shed from eaves and adjacent to an old wall on a shallow or non-existent foundation has often hastened collapse. Problems can be resolved sometimes by underpinning the plinth in short sections or, using suitable shoring and cradling, a leaning wall may be eased back onto a wet mortar bed and new footing using hydraulic jacks.

Problems associated with the rehabilitation of earth buildings

During the rehabilitation of earth walled buildings the suggestion or recommendation of introducing damp coursing is sometimes made. This should not occur before all sources of damp have been correctly identified and eliminated. Correct drainage around an earth building is particularly important.

The introduction of any kind of horizontal damp course is likely to be quite undesirable in terms of physical disturbance and, if effective, in terms of subsequent damage following the excessive reduction of moisture content in the earth wall. Earth walls should not be allowed to dry out completely. Clay is only able to act as a binder in earth walling while it is damp. When it has become very dry it is easily damaged by impact and abrasion and the surface is easily washed away when water has access to the surface. Chalk cob, too, becomes very powdery when completely dry. Moisture contents of 5–8 per cent in such walls are to be expected and accepted.

The only recommended treatment to the wall plinth or pinning is to ensure that it is in good order and to tamp and point as required to maintain it.

Fierce drying, such as that associated with local heat sources, can be particularly detrimental. Fireplaces and hearths are always brick or stone in traditional earth buildings but during modern rehabilitation radiators are placed against earth walls and can create local surface failures. Earth structures which are to be rehabilitated as heated domestic, museum or office accommodation should either be lime plastered, incorporating a haired undercoat, with heat insulation boards behind heat sources, or dry lined with a system incorporating a vapour barrier and permitting ventilation behind the lining. The first approach is usually the simplest, most economic and most acceptable in conservation terms.

7.4 THE REPAIR OF COB, CHALK MUD AND PISE

Repair methods to be avoided

The repair, as distinct from the rebuilding, of chalk mud, cob and pise presents a particular problem because many of the original construction techniques such as treading, and hence compaction, cannot be used to form patches or insertions.

Up to now repairs have tended, therefore, to be mortar repairs or insertions of brick (often calcium-silicate brick in chalk cob) or concrete blocks which are subsequently painted.

- The brick/block method of repair should be discouraged in almost every situation. These materials are so dense and strong that they almost inevitably create perimeter problems due to accumulation of moisture and different thermal properties.
- Mortar 'repairs' usually create problems by inhibiting evaporation from the wall and setting up dam effects because they are almost invariably dense cement or cement-gauged rendered patches.

Both systems of repair are extremely unsightly.

Recommended repair methods for chalk mud and pise

Chalk mud and pise are capable of being repaired by cutting out, mortar repair and stitching. Repair methods, materials and mixes must respect that these materials are normally adequate in compression (unless they are excessively damp), but very weak in tension. The methods described below are generally recommended for in situ repair, although local variations in repair mixes will be considerable.

'Reconstituted' mixes

Chalk mud/chalk cob

Where original material has fallen and is lying on site, or where it is cut out for repair, it should be sieved and treated with a suitable biocide. Fallen material will be deficient in binder. If insufficient material is available, the possibility of digging for additional material on site should be investigated. The raw material will almost invariably be immediately local.

The mixes shown below are typical of successful repair mortars arranged for exposures of increasing severity. The selected sand should match the earth in

Mix A

Non-hydraulic lime putty : Cob : Sand
1 : 3 : 1

Mix B

Gauge 10 parts of A with 1 part white cement *

Mix C (most severe exposure)

Hydraulic lime* : Cob : Sand
1 : 4 : 1–2

*Hydraulic additives are used to provide a small amount of set to compensate for the weather-ability which would have been provided during the original construction by compaction

colour or be sufficiently pale in colour not to influence the final appearance. A weak binder is needed which will also act as a good plasticizer. Lime putty is used to provide this.

Reinforcement, usually short (150 mm) chopped straw should be added in the rough proportion of 2 buckets to 1 m^3 of mix, should the original mix require it. The best way to mix in the straw uniformly is to spread out the mortar, scatter some straw on top and fold in the mortar, repeating this procedure until all the straw is combined.

The above 'reconstituted' cob/pise mortars are extremely stiff to work but should be rammed, chopped and beaten by hand. It may be tempting to use a mechanical mixer if large quantities are required but with the correct, that is, low water ratio needed it is difficult to mix successfully in this way. The danger lies in adding water to increase plasticity; in no circumstances should this happen. Ideally for mixes A and B almost all the water should come from the lime putty. (The lime should preferably be slaked and stored for at least a week; alternatively hydrated lime run to a putty and left to soak for a minimum of twenty-four hours may be used. See Volume 3, Chapter 1, 'Non-hydraulic lime'.) The mixed material should be very sticky, hanging on the shovel. Mixes using hydraulic lime or white cement should be used as soon as possible after mixing. Mix A can be stored in bins or otherwise kept from the air for as long as required, provided that it is thoroughly 'knocked up' just before use.

Clay cob

Clay cob may be reconstituted following similar procedures but using mixes such as:

Mix A

Non-hydraulic lime putty	:	clay cob	:	sharp sand	:	PFA*
1		10		2		1

Mix B

(stronger mix)

Hydraulic lime*	:	clay cob	:	sharp sand
$1\frac{1}{2}$		10		$1\frac{1}{2}$

*See previous note on hydraulic additives

Repair of deteriorated faces

Cutting out

An area should be marked out with a rule and the edge of a trowel enclosing the failed area in a regular frame. A sharp-bladed knife is next drawn along the rule to cut into the surface of the cob. Where the blade strikes large stones it should be withdrawn, commencing the other side of the obstruction. The procedure from here will vary according to the surface area of the repair as follows:

- 'Small' surface areas, that is up to 100 mm square: Use the knife to cut back 50 mm to a clean-backed cavity
- 'Larger' surface areas, that is up to 500 mm square: As above but cut back 100 mm
- 'Large' surface areas, that is over 500 mm square: Cutting back 100 mm must be followed by installing some support for the repair. The height of an unsupported repair should not exceed 500 mm. Support can conveniently be introduced in the form of clay roofing tiles (plain tiles). To insert these, a slot is first cut in the cob to about half the length of the tiles in depth, using a hand saw. The slot is likely to be 125 mm deep and somewhat ragged because of the aggregate that pulls out, but it must be kept to a level line. The tiles will be seated in the slot and protrude to the line of the wall face. The tiles must be soaked in water and the slot must be well-wetted (but not saturated) with water from a hose and fine spray attachment or a hand spray. The slot is filled two-thirds full with the sticky repair mortar and the ends of the tiles bedded firmly in, leaving a 100 mm cantilevered shelf projecting from the back of the repair. The tile shelf/shelves are intended to take the weight of the repair, to provide a barrier which can be compacted onto and to bond the repair into the existing wall. Tile shelves should be set in at 'course' lines where these are part of the original construction and are identifiable. Otherwise the 500 mm rule applies.

  An alternative would be to substitute lengths of expanded stainless steel mesh instead of tiles, but inserted in the same manner. The bottom of the repair must have a horizontal surface and should be formed up and filled first. The mesh should then be inserted for the full length of the repair and the next lift filled
- 'Large' surface areas, where they are to be rebuilt off the pinning may also be built up in courses of clay tiles in a weak mortar (lime putty: sand 1:4) and rendered in a mortar with the addition of hair or synthetic fibre as reinforcement (see page 107, 'Rendering and plastering').

Filling in

The back and sides of a repair area must be well wetted but not soaked. Because the edges are very delicate a weak acrylic emulsion solution (1:10 in water) may be sprayed on to impart a slight strengthening to the cob and to reduce suction on the repair mix. Clay-containing soils should only be very lightly sprayed because of the severe reduction in compressive and adhesive strength that water can cause in this material.

It is not practicable to build up cob in thicknesses of less than 50 mm, and it can be placed for the full depth of repairs of 100 mm. Temporary shuttering enables the mix to be packed down firmly, although the last void must be filled by applying the mix from the face. When the shuttering is removed the face of the cob may be lightly scraped down with the edge of a hacksaw blade to remove the shutter marks.

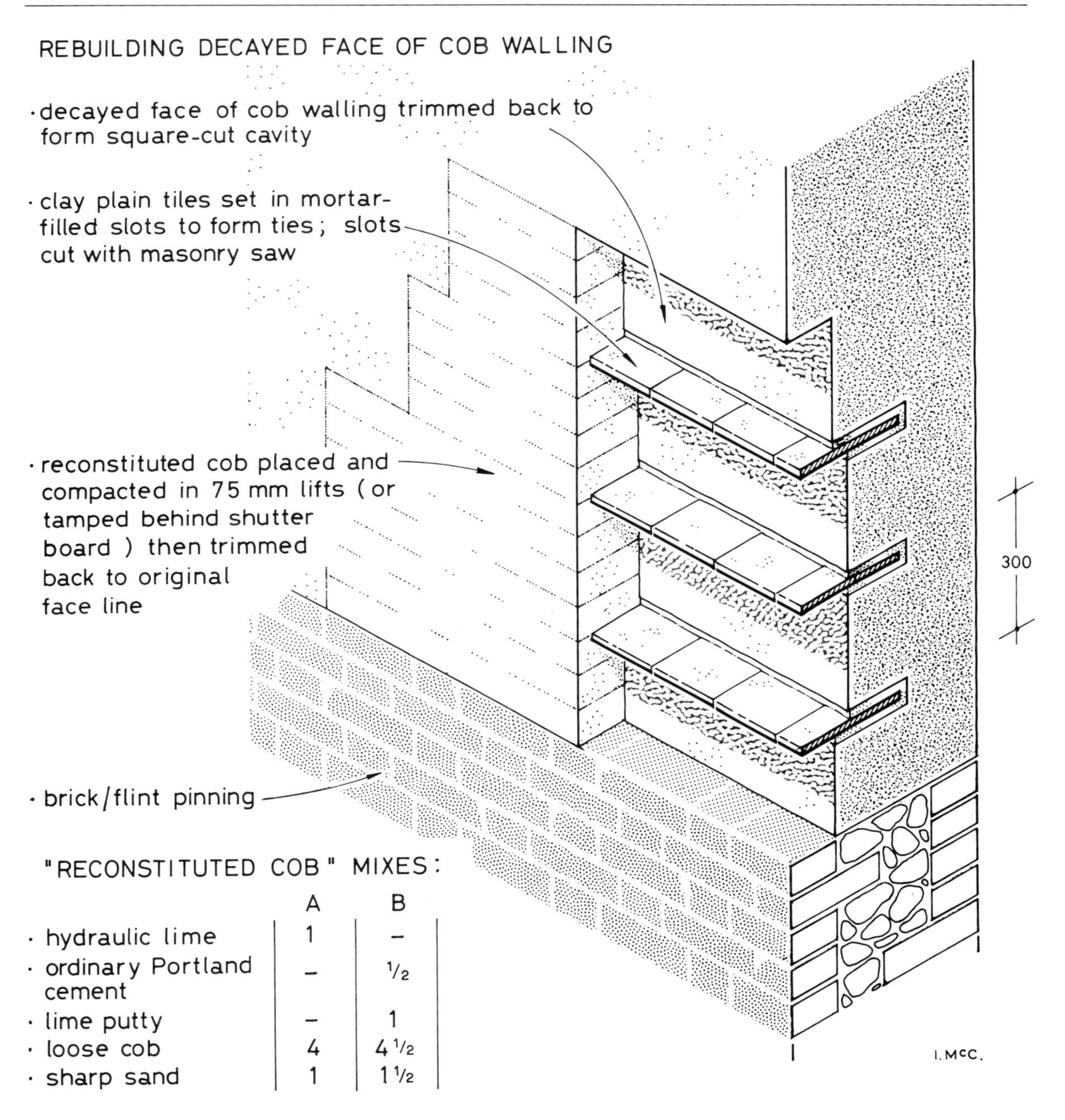

	A	B
· hydraulic lime	1	–
· ordinary Portland cement	–	1/2
· lime putty	–	1
· loose cob	4	4 1/2
· sharp sand	1	1 1/2

Figure 7.2 Method of rebuilding a cob wall face

The failed area of cob (chalk mud) wall shown in this illustration is typical of a splash zone where the surrounding soil has built up too close to the top of the brick pinning. Soil failures can lead to the undermining of the wall and encourage collapse. A 'reconstituted cob' mix including hydraulic lime and sieved cob material is being compacted behind a temporary shutter on to a projecting course of plain clay tiles. When this mix has set and the shutter removed the areas below and above will be filled in with a similar mix. The tile coursing ties the new work into the old.

Cracks

Cracks will develop in cob and pise walls for a variety of reasons, including settlement of footings, decay of timber over openings or injudicious enlarging of openings. All may be enlarged or initially encouraged by saturation through water penetration. Rendering over such cracks with impermeable plaster will further accelerate deterioration by holding water in the vulnerable area. Repairs can be undertaken by stitching and grouting the crack (see 'Stitching, bonding and grouting', below).

Settlement at wall base

If the footing of brick and/or stone has settled and fractured it may be possible to insert light underpinning and simply grout the fracture. However, if settlement is more serious and the cob/pise has fractured and bulged well out of line there will be no option but to take down and rebuild the affected section ensuring that the new base is adequately constructed on a well rammed base.

Temporary lightweight shoring against boards on the wall face may be a necessary first-aid/safety part of the operation but traditional 'remedial' work in the form of brick buttressing or iron 'S' or 'X' face plates and rods must be avoided because their bearing area is usually too small. Where such features already exist it may well be necessary to remove them as part of the remedial work, although not, of course, without carrying out additional alternative repairs.

Loss of bearing for lintels

Jambs to openings are frequently damaged and heavily weathered, reducing the original bearing for the lintels. Alternatively, inadequate original bearing or an enlarged opening with an unsuitably heavy lintel may have induced failure. The usual answer to this problem is to rebuild the jambs as brick or block piers which may or may not be lime-rendered and lime-washed. In the case of inadequate bearing this may well be the best solution and its visual success will be determined by sympathetic treatment and practical expertise. However, if the bearing is still sufficient the jambs may be rebuilt behind lifts of shuttering as described above, or slightly oversized blocks of reconstituted 'cob/pise' may be cast and laid in a lime slurry, paring down afterwards with a saw blade.

Stitching, bonding and grouting

When a section of wall is rebuilt or extended it is recommended that an undercut is made into the old wall at about 60° rather than a straight joint. This provides an inclined area of original wall onto which the new cob can be compacted.

New work may be bonded to old and fractures may be stitched across using stainless steel expanded metal or brass gauze in strips of about 100 mm width bedded in one of the mortar mixes already described. Ties for new work are set into saw cuts in the existing cob/pise in the hydraulic lime mix. At least 300 mm of reinforcement needs to be bedded into the existing and the new work at approximately 0.5 m intervals.

A method of stitching across a fracture in a cob/pise wall is illustrated in Figure 7.3. The gauze or mesh strip is bedded on edge. A chase is cut across the fracture,

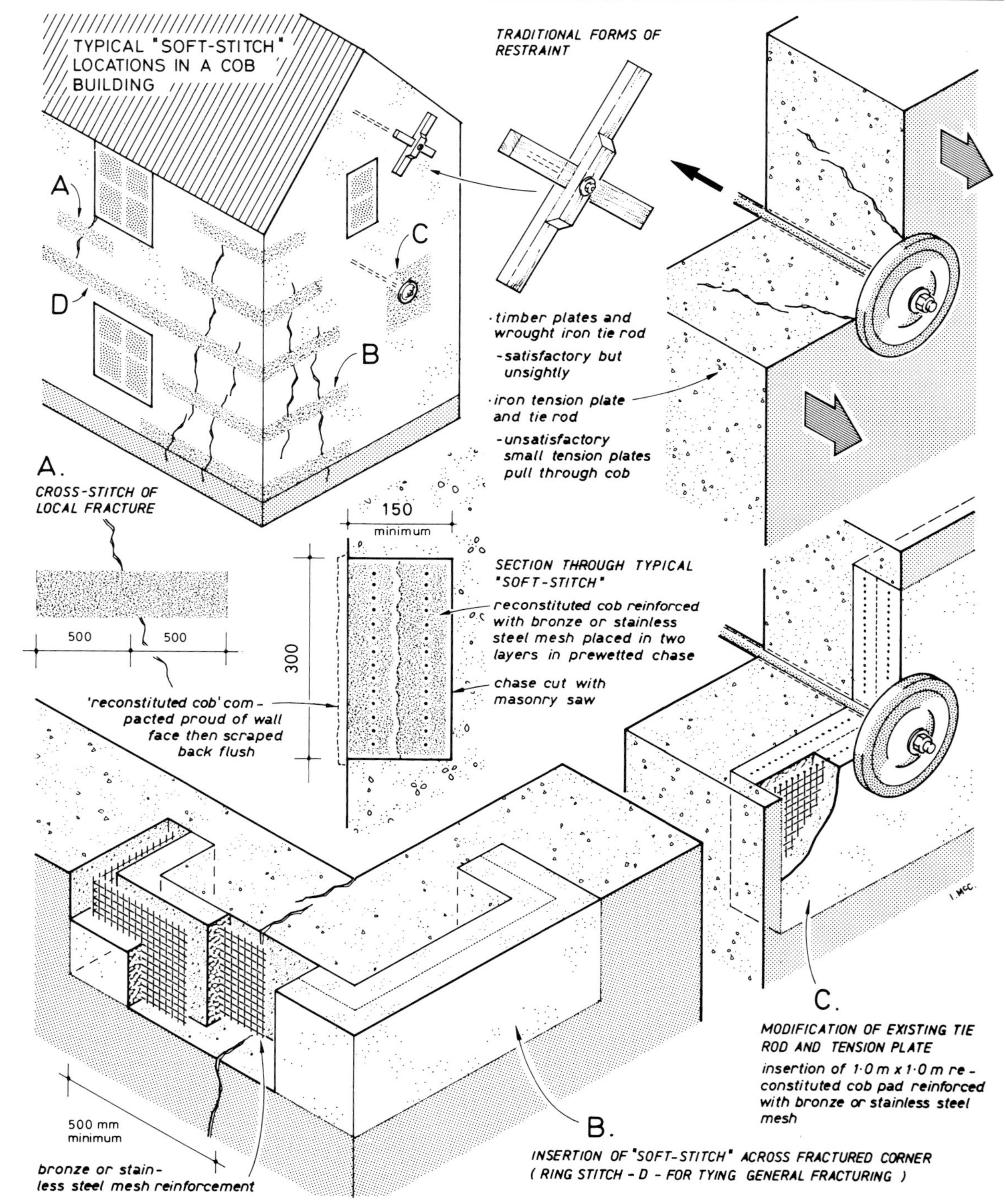

Figure 7.3 Methods of stitching and tying cob walls

using a masonry saw. The chase should be cut as cleanly as possible. The gauze or wire strip is cut to fit the length of the chase. The chase is wetted and then one-third filled with mortar. The strip is pressed home into the mortar and the chase flush filled. These reinforced mortar stitches should be set across large fractures at approximately 1 metre centres. Where a fracture extends right through a wall, stitches should be made to both sides of the wall. The length of the stitch must be decided to suit every situation, but it is unlikely that anything less than 900 mm would be worth inserting. Tension rods and plates can be used in conjunction with these stitches.

After stitching the fracture should be grouted with a liquid mortar. Small fractures or excavations made by animals, such as rat runs, should be filled in the same manner, without stitching. Having been spray-wetted the surface of the crack is first plugged with mortar placed by hand. When the mortar has cured the fracture must be well flushed and then filled by hand pouring with hydraulic lime to which some finely powdered chalk or brick dust has been added; for example, 1 part hydraulic lime: 3 parts chalk dust or brick dust (to pass 600 micron sieve), material selected to suit the colour of the wall. The quantity of water should be enough to create a very mobile grout. Ten per cent acrylic emulsion in the gauging water will assist adhesion to the fracture and aid initial mobility.

Rendering and plastering

There will often be an external render to a cob wall. In buildings there may also be an internal plaster. Whatever the exposure it must be remembered that the cob/pise has a relatively weak surface; it cannot tolerate strong renderings. A lime:sand composition of 1:2½–3 prepared from lime putty floated firmly onto a well dampened surface is about the strongest composition that should be applied inside or out. (See also Volume 3, Chapter 4, 'External renders'.) A weakened surface may be assisted by the inclusion of an alkali-resistant lightweight fabric. A wash comprised of lime putty and fresh cow dung (in equal proportions and left to stand for at least a week after mixing), or whiting and dung mixed similarly, may be used in place of the usual lime-tallow wash where a beige colour is acceptable, although it is less durable.

Limewashing

Limewash is often the first line of defence that an earth wall has towards the weather and the delicate nature of earth wall materials makes the maintenance of a continuous coating imperative. The preparation of traditional limewash is described in detail in Volume 3, Chapter 7, 'Limewashes and lime paints'.

When limewash is applied to a clay-based earth wall, the first coat will combine with the soil to form an earthy slurry which will crack as it dries out. This is to be expected. Two more coats of limewash should be applied and these will hide the cracks and provide solid colour.

Consolidant for clay earth walls

Clay walling has sometimes been effectively consolidated by spraying ethyl silicate ('silicon ester'). This reacts with the clay particles forming a tridimensional network of silica bridges which increase the water resistance of the material. The surface maintains its original porosity, with the advantage that the internal moisture can evaporate, and that further treatments of any kind can be performed in the future. There is, however, the possibility of soluble salts building up behind the consolidated zone. Silicon esters do not have gluing properties, and if pieces of mud wall are already detached they cannot be held together. Consolidation should only occur after all water penetration has been dealt with and if there is no soluble salt problem present. Although there is no current experience known on the use of other alkoxysilanes in earth walls, there is no reason to suppose they would not be equally effective.

7.5 REPAIR OF CLAY LUMP

Clay lump construction is a true masonry construction, unlike chalk mud and pise. For this reason, although local stitching or surface patching may be appropriate methods of repair as described, it may also be possible to form new clay blocks and to cut out and piece in the traditional masonry manner.

'Reconstituted' mixes: patch repairs or stitch filling

If a patch repair or stitch filling technique is to be used as for chalk mud or pise, as much fallen soil should be salvaged as possible. As the finest component has probably been washed away, this may need to be supplemented with freshly dug clay or soil from the locality. Spread the soils out on a boarded base and allow them partially to dry out so that they can be broken into small pieces with a rake or hoe. Sieve the soil through a 12 mm sieve and add hydrated lime, pozzolanic fly ash (pfa) and sharp sand as follows:

Parts by volume				
Dried clay-based soil (well graded)	10	*or*	Dried clay-based soil (well graded)	10
Hydrated lime	3		Hydraulic hydrated lime	$1\frac{1}{2}$
PFA	1		Sharp sand	2
Sharp sand	2			

The sand should be mixed with the clay soil first, followed by the dry lime (and fly ash where appropriate). When a good dry mix has been achieved, spread the material out, sprinkle with water, fold together and repeat until a stiff, very sticky mortar has been formed. Keep the water ratio as low as possible. Although this is a laborious operation additional water must not be added simply to make the work easier. Only a little water is required to make clay workable – good mixing is required to make sure close contact is made between the clay and the water. Unfortunately, mechanical mixers are not very practicable for such a sticky material and there is no satisfactory alternative to hand mixing.

Preparation of the wall, placing, reinforcing and finishing can proceed as for chalk mud and pise.

Cutting and piecing in

Where possible, deteriorated clay lump should be sawn out to the old joints for the full depth, providing the necessary supports as work proceeds, and pieced in with new clay blocks. The method of manufacture should follow the general procedures and guidelines set out below.

The manufacture of new blocks

Once the soil composition of the existing blocks has been determined and matched production of new blocks can begin. The soil must be dug from the subsoil, all stones, gravel and organic material removed, and then screened through a rectangular wooden frame with 6–12 mm mesh. An optimum, minimum amount of water should be added depending on the proportions of clay and silt in the soil. Place the soil in a wheelbarrow and mix it with a hoe, gradually adding small amounts of water. Enough water has usually been added once the soil begins to 'ball' or roll down the steep slope of the barrow. Straw should be added at this stage if it is part of the original mix.

Determining the mortar proportions

The mortar material between clay lump blocks should not be the same as the blocks as the joints should be the weakest zones along which fissures appear when the masonry is loaded. The mortar must be tailored to the absorption of the adjacent blocks.

To know if the mortar contains the best proportions of clay binder and sand stabilizer a series of clay lump sandwiches with mortars of varying percentages of sand and clay should be made. Two blocks should be buttered together with the test mortar to form a joint 3 mm thick and opened the following day. If no visible fissures are found, the sand/clay proportion is close to the optimum for the blocks of that wall.

Replacing blocks

When stitching across cracks or replacing deteriorated blocks, whole blocks should be removed to form a flat bed along the opening. All vertical and horizontal joints should be lightly dampened then filled with mortar to ensure a tight fit. Only very light, if any, water spraying should occur before the mortar is placed.

Cracks in clay lump walls

As in other masonry constructions, cracks may run through joints and across blocks. Cracks may be grouted, much as described before, using hydraulic lime, brick dust and water gauged with acrylic emulsion. It is not recommended that crushed and sieved clay is used in the grout, so that a suitable mix might be 1 hydraulic lime to 3 brick dust, with 10 per cent acrylic emulsion in the gauging water. Major fractures through blocks may be stopped up with a weak lime mortar as a temporary measure, but the proper solution is to cut out and replace. A temporary support such as a stainless steel flat should be let in above the defective block, spanning the area to be cut out with at least 100 mm bearing each side. The slot for the steel is sawn out with a masonry saw. Carefully saw and cut out the defective block, material from which may be used to reconstitute the new block. Wet the clean cavity, butter the beds and sides of the block and ease into position. When placed, the flat must be removed and the slot stopped with weak mortar.

Rendering, plastering and limewashing

As for cob and pise.

Tarring

Traditional coal tar is no longer available. Modern bituminous paints will not provide as thick a coating as coal tar did, but they are the most appropriate alternative. More than one coat will be required to build up thickness. New bituminous paint may initially have a shiny surface.

REFERENCES

1 Alva Balderrama, Alejandro and Chiari, Giacomo, *Protection and Conservation of Excavated Structures of Mudbrick*, ICCROM unpublished paper, Rome, 1982.
2 Alva Balderrama, Alejandro and Teutonico, Jeanne Marie, *Notes on the Manufacture of Adobe Blocks for the Restoration of Earthern Architecture*, unpublished paper, ICCROM, Rome, 1983.
3 British Standards Institution, *BS1377: 1975 Methods of Test for Soil for Civil Engineering Purposes*.
4 Brunskill, R W, *Illustrated Handbook of Vernacular Architecture*.
5 Clifton-Taylor, Alec, *The Pattern of English Building*.
6 Department of Housing and Urban Development, *Handbook for Building Homes of Earth*, Office of International Affairs, Washington DC.
7 Harrison, J R, *The Mud Wall in England at the Close of the Vernacular Era*, 'Transactions of the Ancient Monuments Society', Vol 28, 1984, pp 152–172.
8 Harrison, J R, *Traditional Cob and Chalk Mud Building in the UK and Eire, A General Survey*, unpublished dissertation, Diploma in Conservation Studies, Institute of Advanced Architectural Studies, York, 1979.
9 McCann, John, *Clay and Cob Buildings*, Shire Album 105, Shire Publications Limited, 1983.
10 Middleton, G F, *Earth-Wall Construction*, Bulletin No. 5, Department of Transport and Construction, Experimental Building Station (NSW), Australian Government Publishing Service, Canberra, 1982.
11 Pearson, Gordon T, *Chalk: Its Use as a Structural Building Material in the County of Hampshire*, unpublished thesis, Postgraduate Diploma in Building Conservation, Architectural Association School of Architecture, April, 1982.
12 Torraca, Giorgio, *Brick, Adobe, Stone and Architectural Ceramics: Deterioration Pro-*

cesses and Conservation Practices, 'Preservation and Conservation Principles and Practices', proceedings of the North American International Regional Conference, Williamsburg, Virginia, 1972, The Preservation Press, 1976, pp 143–165.

13 Williams-Ellis, C and Eastwick-Field, J E, *Cottage Building in Cob, Pise, Chalk and Clay*, 1920 and 1947.

14 Williams-Ellis, C, 'Cottage Building in Cob, Pise, Chalk and Clay', in *Country Life*, 1919, page 104.

See also the Technical Bibliography, Volume 5.

8 FLOORS OF EARTH, LIME AND GYPSUM

8.1 INTRODUCTION

The ground floors of many ancient buildings were frequently no more than well compacted earth, over which straw or rushes might be laid as a renewable covering. Their quality and survival depended on the nature of the soil, the drainage of the site, the maintenance of a roof and the extent and type of use. When lime or gypsum was included a type of weak concrete slab or screed could be formed, these were used on ground and upper floors.

Where possible and where the original floor material is known, repair or reinstatement should be of similar type. Modifications may be necessary to meet the demands made on the floor by hundreds of visitors, but primitive flooring materials can be remarkably resistant to wear when well prepared. Earth floors may need some assistance at thresholds in the form of open weave rope mats or small purpose-made duck-boards during wet seasons if there are many visitors.

There are many different types and variations so that the remedial work on old floors of this kind must only take place after a close examination of what survives. When disturbance of old floors takes place there must always be archaeological supervision.

In this chapter typical repair or replacement guidelines are given for earth, lime and lime-gypsum floors.

8.2 EARTH FLOORS WITH CLAY

Surviving earth floors which require stabilization may be treated simply by rotivating the top surface, raking in up to 10 to 15 per cent of the soil volume of hydrated lime, watering by rose sprinkler and compacting with a vibrating roller. This is only likely to be effective where the soil has a natural clay content.

To determine whether or not clay is present, take a representative handful of unweathered soil, dissaggregate it, place it in a tall, straight-sided jar, add enough water to form a slurry, break up any remaining lumps, then fill the jar with water and shake vigorously. Let the jar stand overnight or until the water is clear, then syphon off the water without disturbing the upper layer sediments. Any clay will have settled on the top of the deposited sediments and, once these have been partially dried, will be plastic and greasy to the touch. A rough proportion of the clay to other constituents can be seen through the side of the jar.

8.3 EARTH FLOORS WITH LITTLE OR NO CLAY

If the simple test shows little or no clay, stabilization can be achieved by rotivating and raking in 10 to 15 per cent of cement made up of equal parts of ordinary Portland cement, lime and flyash (pulverized fuel ash), raked or rotivated, watered and compacted as above. Proper preparation and compaction should give good results. Again, excessive wear at thresholds can be a problem, and some form of duck-board mat may be necessary. The composite cement proportion may be increased to resist anticipated heavy wear, but the character of the floor will begin to change accordingly.

Local areas can be cut out to 150 mm depth with a sharp turf cutter or a masonry saw before repair, but it is better to treat an entire floor at one time if possible.

8.4 EARTH FLOORS WITH OR WITHOUT CLAY

A stronger, more resistant floor than 8.2 and a rather weaker, less resistant floor than 8.3 can be formed by using 10 to 15 per cent of fresh hydraulic lime as the added binder in place of lime, cement or PFA.

8.5 LIME CONCRETE

Lime concrete was used extensively by Roman and later builders using crushed brick aggregate and a cementitious matrix of lime putty mixed with brick dust or volcanic ash. Massive structural concrete and void filler frequently incorporated lightweight tufa or pumice aggregate. Pozzolanic concretes for floors and fill are conveniently matched using the same materials as these, or using PFA furnace bottom ash as the pozzolanic additive. Examples of suitable mixes, by volume, are:

- Lime : PFA furnace bottom ash : Sharp sand
  2 : 4 : 1
- White cement : Lime : Brick dust/lightweight appregate
  1 : 2½ : 5

8.6 BLOOD FLOORS AND SCREEDS

The addition of fresh blood to earth, earth-lime or lime concrete floors creates a hard-wearing surface which can take a certain amount of polish. It is, of course, unlikely that such a floor would be laid today, but there are many traditional precedents and it may be thought appropriate in the case of a specific site.

Fresh blood is best used with a lime or hydraulic lime binder as a screed. Mix the hydrated lime with $2\frac{1}{2}$ parts (by volume), well graded sharp sand and grit. Water until damp, but not yet plastic. Add fresh blood obtained from a local abattoir, slowly, turning over the mix until it is fat and easily workable, but not too wet. Wet the substrate so that it does not take moisture for the screed and tamp the blood mortar. Level firmly and uniformly with a board. Finish the screed by working it with a wood float. If a polish is required this can be formed by alternatively scouring and floating with a little water as the set begins to take.

Because there is a risk of mould growth the entire area should be spray treated with a biocide when cured. Needless to say this type of floor is not recommended for regular living accommodation. Rubber gloves should be worn and a simple gauze mask may be desirable.

8.7 LIME-ASH FLOORS

Traditional lime-burning leaves a residue of quicklime, unburnt lime and coal or wood ash which has, at times, been used for forming floors or for rendering infills in timber frame buildings. The following two traditional specifications are examples of ways in which lime ash has been used in earth floors.

Northampton and Nottinghamshire tradition

'One-third lime, one-third well sifted coal ashes and one-third loamy clay and horse dung made from grasses'.

Mix materials in a dry place, let it stand for ten days then mix again, adding a small quantity of water, let it stand once more for three to four days, mix once more, then the mix is ready to be laid. The mix is laid to the required thickness by treading and beating with a hammer. On an upper floor, the trowel only must be used and the mix allowed to dry slowly, the mix being bedded upon reed or lath between it and the joists.

Hardwick mix

The mix is by volume 4:1 lime-ash:plaster of Paris (Class A plaster).

The lime-ash is first wetted and mixed and is then opened up and more water added. The gypsum is then puddled in and the whole properly mixed. Because of the relatively rapid set of the gypsum, care must be taken that no more of the mix is made than can readily be used at once. Depending on the area involved, a screeding board may be used. The mix should be well worked in but not over trowelled, as staining from the ash can result. The thickness of the flooring varies between 75–85 mm laid in two layers, the top layer being wetter than the first.

This mix has been used with ideal results at Hardwick Hall, Derbyshire, where the floor construction is joists, laths and a thin layer of hay.

PFA substitutes

An alternative to lime-ash flooring is a mix of pozzolanic PFA and ordinary Portland cement to make an 'all fines' concrete. A suitable mix would be 1 part OPC:3–4 parts PFA laid 150 mm (6 in) thick. White cement and lightweight aggregates may also be used, with or without lime on suspended floors. Remember that, even if a gypsum floor is being replaced, cement and gypsum must not be used together.

9 THE EXAMINATION, ANALYSIS, REPAIR AND REPLACEMENT OF DAUB

9.1 DAUB PANELS AND TIMBER FRAME CONSTRUCTION

Daub construction consists of an application of earth-based plaster (daub) to a backing of woven hazel twigs or riven oak or chestnut lath. Daub panels were a common method of infilling traditional timber frame buildings, the dominant form of construction in England and Wales prior to the seventeenth century, and common construction in some areas until the late eighteenth century. Daub panels were used in internal and external walls (reference 3).

Local and regional variations

Because daub panels were a vernacular form of construction, variation should always be expected in soil type, mix proportions, thickness of the panels and in the number of layers of application of the earth. Close inspection of the case at hand is of vital importance to determination of the correct solution to problems of repair and replacement.

Wattle and daub, lath and daub

In wattle and daub, holes or slots about 1 in (25 mm) in diameter are driven in the underside of the timber member forming the top of the panel, and a groove is cut in the base member. Oak staves with pointed tops and chiselled bottom ends are positioned in these about 12–18 in (300–450 mm) apart. They form the strong basis around which wattles, usually of hazel or cleft oak, are woven basket-fashion. Alternatively, the backing was formed by riven oak laths sprung in or, later, fixed with wrought iron nails, horizontally between studs and framing. Approximately $\frac{1}{2}$–$\frac{3}{4}$ in (12–18 mm) was left between the laths to provide a key for the plaster. The wattle or lath panel was daubed from both sides. The completed panel was then limewashed.

Clay-earth daub mixes

There are many traditional mixes for daub plasters. Mixtures of soil, generally medium to fine grained, dung and chopped straw or hay were common. A small proportion of clay and roughly equal proportions of silt and sand were often present in the soils used. Crushed chalk or other, larger gravel aggregate was also used at times. Old or weathered dung, usually cow dung, was introduced to improve workability and durability. Shrinkage was controlled by the sand and other aggregates present, chopped straw in 4–6 in (100–150 mm) lengths, and the use of a mix which was as dry as possible. Lime was rarely used to stabilize these mixes. A measure of salt was at times introduced to assist the daub to retain a small amount of moisture and hence not dry out completely. This served to control shrinkage cracking.

Details of the contents of several clay/earth daubs can be found in section 9.4 of this chapter, 'The analysis of daub'.

Lime-gauged daubs

Lime-gauged daubs are also known, and again there are many traditional mixes. A typical serviceable one is 4 parts lime putty:1–1½ parts sharp sand:1 part slurried cow dung, reinforced with 6 in (150 mm) chopped straw, thoroughly mixed, beaten and rammed. The mix which is used as dry as possible, may be scratch-keyed and finished with the 4:1:1 mix, sieved and without straw.

9.2 DETERIORATION OF DAUB

If properly maintained, daub panels will last indefinitely. As with all forms of construction involving unfired earth, water must not be allowed to penetrate the protective outer surface otherwise rapid decay takes place. When, unfortunately, this did occur, many daub panels were replaced with brick or, more recently, blocks or plaster boards. Timber frames often became distorted by the insertion of heavy, rigid substitute. Essential maintenance of daub should include annual inspection and possibly annual renewal or touching up of external limewash.

Hazel twigs in wattle and daub panels were selected primarily for their flexibility and length, not their natural durability. These were often attacked by common furniture beetle (anobium punctatum) or by lyctus powder post beetles, and rendered structurally ineffective. Where hazel twigs are woven around oak staves this is usually not a problem as the daub and the uprights can be sufficiently integrated to continue to act as a stable panel.

Wattle and daub and lath and daub panels are particularly resilient to impact. While the daub, which is very weak in tension, will crack and at times lose key, the wooden network will usually remain undamaged. Daub may also crack due to decay or distortion of the supporting frame.

The presence of soluble salts within the daub is not normally a serious problem but can result in powdering surfaces and breakdown of limewashes. Limewash is, however, the only paint which should be used (Volume 3, Chapter 7, 'Limewashes and lime paints'). Panels which include dung must not be covered by

Replacement of traditional daub in a timber-framed building onto riven oak lath. The daub was based on a well-graded earth containing about 15 per cent clay, with the addition of animal dung and straw. The overall shrinkage of this panel is very low and the surface cracking within acceptable limits. These daubs were protected with lime, or a lime clay plaster, or further daub, or, as would be in the case shown, three coats of limewash. The first coat would be well-worked into the surface, filling the small cracks with the slurry so formed.

impervious paint or rendering because the action of anaerobic bacteria will commence within the dung, weakening and breaking down the daub.

9.3 REPAIR AND REPLACEMENT OF DAUB

Repairs to damaged or deteriorated daub panels must recognize the simple and sensitive nature of this form of construction. It is essential that original materials, mixes and construction techniques are repeated as far as possible.

RTAS Daub Research Project

During 1985 the Research and Technical Advisory Service carried out a research project and associated fieldwork into the analysis and replacement of daub. The work was carried out on the North Cray Medieval House at the Weald and Downland Open Air Museum, Singleton, West Sussex. Samples of a single thickness daub had been analysed (see section 9.4) and several panels were recreated.

Replacement of daub

While every attempt should be made to repair daub panels, it will at times be necessary to replace the backing and/or the daub plaster.

Inspection and recording of existing panel

Before and during removal of the deteriorated panel, close observations and records should be made of details of construction, such as spacing of laths, type of fixings, width of the woven hazel and position of the face of the finished daub. The nature of the finished surface and any coating and the constituents of its mix must be determined. The number of coats, the existence of cross-keying and edge filling should also be identified. Not all backing may need replacing, and some elements such as wrought iron nails may be re-used.

Replacement of backings

New wattles or riven laths may be required. It is still possible to purchase these in the UK. A new wattle panel should be woven lightly and once seasoned, firmly compacted from the bottom up and additional wattles inserted at the top.

Analysing the daub

The constituents of *clay-earth* daub may be determined from careful visual inspection and from sedimentation analysis. Section 9.4 describes the application of these methods. It is necessary to know the proportions of clay, silt, sand and other larger aggregates in the soil, the proportion and type of dung and the amount and lengths of straw. The subsoil in the immediate vicinity should also be analysed to determine whether it will be a suitable replacement soil as it is or whether adjustments such as removal of larger aggregate or addition of sand will need to be made. It should be remembered that sand and straw are the main stabilizing components which will control shrinkage and cracking.

The following mix was used by the Research and Technical Advisory Service during research work on daub repair at the Weald and Downland Open Air Museum, Singleton (1985):

> Based on a soil of 10 per cent clay, 40 per cent silt and 50 per cent sand*:
>
> Use 12 parts soil: 1 part dung: $1\frac{1}{2}$ parts straw (firmly packed). Mix the above with $\frac{1}{4}$–$\frac{1}{2}$ part water
>
> *The constituents of a daub soil will vary from area to area. They can be in the following ranges: 5–15 per cent clay, 20–55 per cent silt, 20–55 per cent sand

Lime-gauged daub will need to be analysed in the same way as a mortar or render, the lime binder being dissolved in acid first. (see Chapter 3)

Mixing the daub

The soil, dung and lime should be mixed first with a minimum of water being added. One-quarter bucket water to 12 buckets soil and dung is probably all that will be needed. If historical precedence shows that salt should be added, this should be dissolved in the water before adding. The ideal way of mixing daub is a pug mill but in the absence of this the materials should be trodden, chopped and retrodden for at least half an hour. Daub cannot be mixed too much. The straw is added towards the end of mixing by sprinkling on top of the daub, folding in the edges and further treading. This process is repeated until all the straw is mixed in evenly and coated with daub. Proper mixing is best achieved by working small batches at a time.

Placing the daub

The following procedure was adopted during the RTAS field work. Once the daub was prepared:

1. The laths, which were very dry, were lightly sprayed with water
2. The daub was applied from both sides and pressed gently but firmly to fill the spaces between the laths. The application began at the bottom of a panel and worked across and up. Although the mix was slightly firm, it was placed with a plastering trowel and finished with a small cross-grained woodfloat
3. For the soil used, to achieve a practical covering without inducing massive shrinkage, it was found that the thickness of daub covering the laths needed to be in the range of $\frac{1}{2}$–1$\frac{1}{4}$in (12-30 mm). This was confirmed by the historic precedence of the original panel thickness
4. The work was undertaken in the summer and each daub panel took three to four weeks to dry properly. This process was carefully controlled. Panels on external walls were protected from direct sunlight and air movement around them restricted. They were covered on both sides with a double layer of hessian sacking which was kept damp during the warmest weather. Panels on internal walls were protected around the edges where the rate of drying was fastest
5. At the end of every day, the panels were checked for cracks. Gentle reworking of recently placed daub eliminated many of these but only at the early drying stages while the mix remained easily mouldable to the fingers. The daub at the edges was gently extended towards the main frame members. Any fine cracks within the panels, typical of daub construction, were eventually bridged with the limewash finish. Gaps between the daub and the frame were pre-wetted slightly, then packed with daub which was as dry as possible, inserted with a clay modelling tool

Repair of cracks and loss of key

Daub panels which are experiencing loss of key will usually bulge and crack. If the wattle or lath construction is still sound the daub coats may be pulled together by inserting non-ferrous surface chases and tightening them together with non-ferrous wire loops, passed through drill holes. If the wattle or lath construction has failed the panel will need to be stripped out and rebuilt.

Repair of patches

Where less than 25 per cent of a daub panel is damaged or deteriorated, the defective area can be repaired in the following way. Cut around the damaged area with a sharp-bladed knife. If possible hold the knife at an angle so that an undercut edge to the repair is formed. Remove daub within this area back to the lath or wattle, gently cleaning around this so that a key is formed for the replacement mix. Lightly spray the area with water, then place the daub mix in one layer, finishing it proud of the original surface. Carefully control the rate of drying by covering it with damp sacking. As shrinkage cracks appear and while the material is still mouldable, gently rework the surface and push the daub towards the edges of the repair patch. Cracks between the existing and repair areas can be filled later.

9.4 THE ANALYSIS OF DAUB: CASE STUDY

Background

Five samples of earth daub were submitted for analysis to RTAS in 1985 by the Weald and Downland Open Air Museum, Singleton, West Sussex. The samples had been taken from buildings rescued by the museum. Analysis was required to determine the constituents of the sample pieces and, subsequently, an appropriate mix based on local soil, for daub panels of a house re-erected at the museum was provided.

General procedures

Each sample was first analysed visually with the assistance of a ×10 magnifying glass. A representative segment of approximately 50 g (2 oz) was then removed, disaggregated and inspected further. These particles were further broken down in water and then subjected to sedimentation analysis to distinguish the proportions of clay, silt, sand, other aggregates and vegetation. Following this drying, sieving and further visual inspection provided additional information concerning additives such as dung, hair, grain and larger aggregates such as chalk.

A second portion of each sample was disaggregated, moistened and worked by hand to assess performance. Drops of 10 per cent hydrochloric acid were placed on a third, dry portion to test for lime binder. At least one-third of the original sample was kept for reference. Details of the analysis procedures used may be found in Chapter 7, Section 7.1, 'Earth walling'.

Results of analyses

Daub 1: Boarhunt, Hants (museum reference: 321/83)
The Boarhunt daub was dull beige in colour with small dispersed orange patches. A few small white lumps were visible. A reasonable amount of straw 50–75 mm (2–3 in) in length could be seen evenly distributed throughout the mix. There were a few shrinkage cracks within the sample.

It was not difficult to break and crumble the sample with fingers. As this daub was further disaggregated in water, the harder, orange clay lumps gradually dissolved, making the water very orange in colour. The orange lumps contained many fine roots. At this stage the white lumps could be identified as chalk aggregate. The sample was thoroughly mixed, shaken and left to settle out in a tall volumetric flask. Floating straw and other vegetable matter was skimmed from the top and retained. After a week the water above the settled layers was clear but stained a tea colour.

These sediments were passed through a 1.18 mm sieve, the vegetation collected and combined with those collected earlier. The constituents were found to be short pieces of straw and grass, seed pods and fine short roots.

A separate portion of this sample was broken up and moistened until mouldable. It was slightly sticky to feel. The damp daub could be formed into a roll 18 mm ($\frac{3}{4}$ in) in diameter before it broke into 50 mm (2 in) lengths which remained connected by the longer straw pieces. This portion of sample was stored in plastic and after one week was supporting a black mould growth.

Conclusions
The Boarhunt daub was comprised of a fine grained soil of approximately 5 per cent clay, 35 per cent silt and 55 per cent fine sand. The clay was probably blended with the silt/sand during the original mixing. The final 5 per cent of the soil was made up of chalk aggregate (3 mm and less). The daub had straw added and contained dung in the approximate proportions – 1 part dung to 9 parts soil/aggregate. The mix did not contain any lime binder.

Daub 2: Walderton Cottage, West Sussex
Visual analysis of the Walderton Cottage daub revealed a beige binder, a high proportion of chalk aggregate 1–5 mm in diameter, a few pieces of flint 1.2 mm in length, a high percentage of short straw and seed pods. One surface was heat affected and blackened by smoke.

Once fractured the sample was easily disaggregated by hand in dry form. Only a small amount of further disaggregation was required under water. Some lumps of binder and small chalk pieces could not be separated. These may have been part of the heat affected and partially vitrified surface which were not discarded from the sample selected.

The water above the strata was stained, very smelly and supporting a mould growth within the floating vegetation. The vegetation collected from the water surface and sifted out of the sedimentation layers was comprised of straw and grass pieces 25 mm (1 in) – 1 mm in length, seed husks and seed hairs.

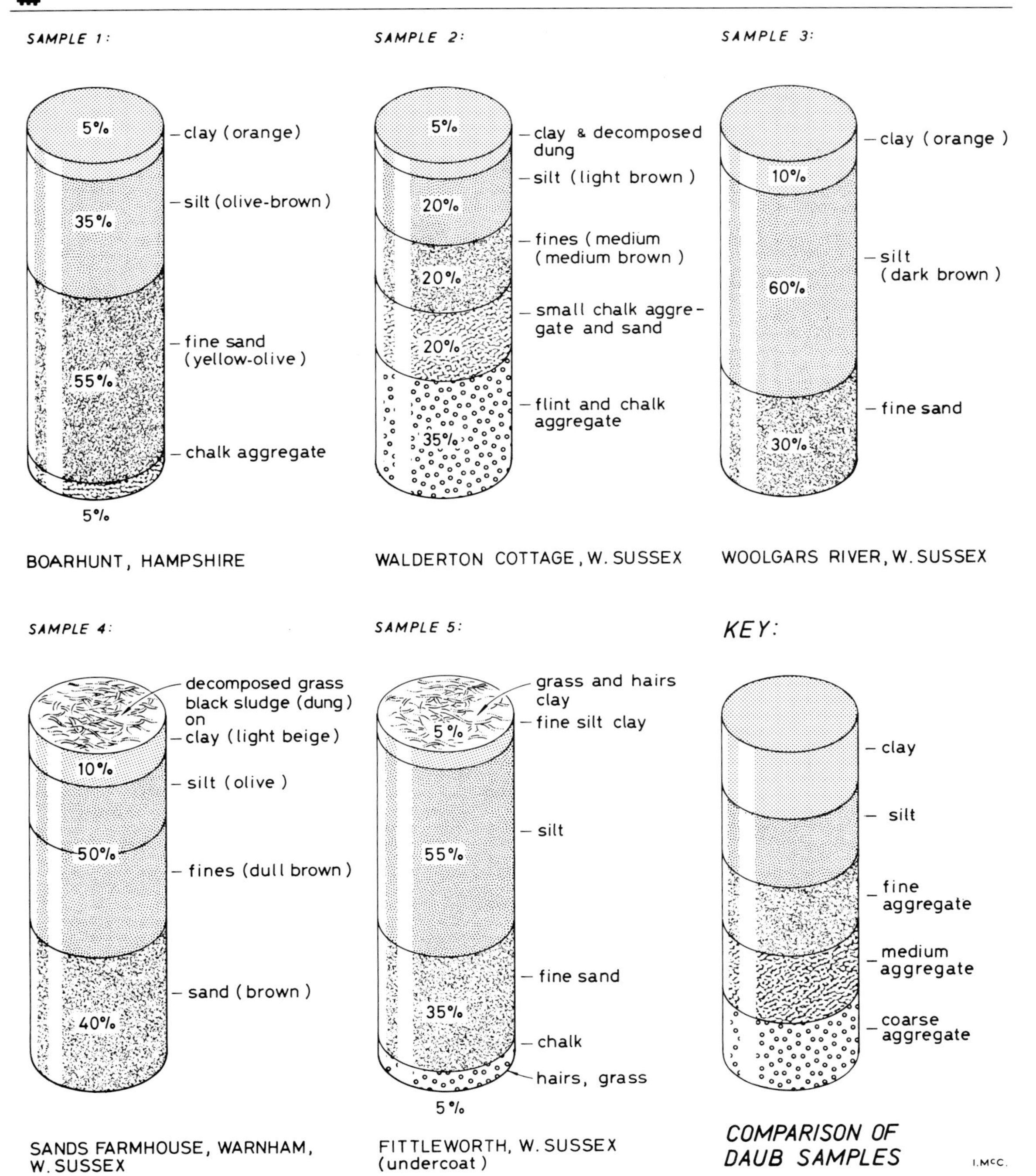

Figure 9.1 Daub samples – sedimentation analyses

The performance test on the second piece of the sample showed that with a minimum amount of water the daub constituents could be made to stick together in a ball. Additional water did not improve adhesion. The dampened materials could not be worked into a roll less than 38 mm ($1\frac{1}{2}$ in) in diameter and 50 mm (2 in) in length without crumbling. After a week under plastic this daub cake was covered in black mould.

Conclusions
The Walderton Cottage daub was comprised of a coarse and well graded soil with a low clay content. The soil was approximately 55 per cent chalk aggregate, 20 per cent sand and fine chalk, 20 per cent silt and 5 per cent clay. The proportion of dung was relatively high at approximately 1 part dung to 7 parts blended soil. Cow dung was probably used. A substantial amount of straw had also been included in the mix, and could have come from dung collected off the byre floor. There was no lime in the binder.

Daub 3: Woolgars River, West Sussex (museum reference: 61/85)
The sample submitted was too small and too extensively heat-affected for a fully representative analysis to be made. It was comprised of a dull beige soil, long pieces of thick, shiny straw (50 mm, 2 in plus) and was extensively coated in black soot. To test for grease, a blackened sample was boiled in water and then cooled. The water remained black and when it had evaporated, a thin tarry residue was left. Inspection of the vegetation revealed pieces of fresh straw about 50 mm (2 in) in length, other shorter straw, some roots, grain and husks.

There was not enough sample for a performance test to be undertaken.

Conclusions
Within the limitations already expressed it is thought that the Woolgars River daub was comprised of a very finely graded soil with a clay content of less than 10 per cent, that the dung soil proportion was approximately 1 part dung:9 parts soil and that a small amount of fresh straw was added or collected with the dung. It did not contain lime in the binder.

Daub 4: Sands Farmhouse, Warnham (museum reference: 322/83)
The Sands Farmhouse daub was comprised of a well-compacted, fine-grained soil mixed with a small amount of straw and husk hairs. Its outer surface had a combed pattern and had possibly been treated with linseed oil to which dirt had become attached. Parts of the sample were smoke-stained and heat affected.

The sample was very hard and had to be disaggregated by crushing with a mortar and pestle. Once the crushed lumps had been in water for several minutes, it was possible to break them down fully. The sedimentation test resulted in very smelly, darkly stained water on top of the soil. The vegetation was mainly small, dark pieces of grass, husk hairs and some 50 mm (2 in) long pieces of fresh straw.

The second portion of sample produced a reasonably sticky, mouldable mix which could be rolled into a sausage of 12 mm ($\frac{1}{2}$ in) diameter. When dry this daub cake was much easier to break than the original sample, probably due to inadequate compaction.

Conclusions

The soil of the Sands Farmhouse daub was finely graded and comprised of about 10 per cent clay, 50 per cent silt and 40 per cent sand. A noticeably high amount of cow dung (1 part dung to 7 parts soil) and some short straw had been added. The sample contained no lime in the binder.

Daub 5: Ivy House, Fittleworth, Sussex (museum reference: 39/76)

The Ivy House daub was a two-coat plaster, beige/orange in colour. The top coat which was 12–15 mm (about ½ in) thick contained a very high proportion of grain and grass seeds and a soft hair in short (12 mm, ½ in) lengths, bound with a fine clayey slurry. The outer surface of this top coat was originally coated in white limewash. A few shell fragments formed part of the aggregate in both layers of plaster. Small pieces of chalk aggregate, a small amount of short hair and digested grasses could be seen in the body mortar.

At the conclusion of a sedimentation test there was no definable boundary between these layers, only a very gradual change from medium-brown at the bottom to a lighter brown at the top.

The reconstituted portion of base coat produced a very sticky, highly mouldable mix. After a week under plastic it did not support a mould growth. When dry it was particularly difficult to break, the hairs being a major contributor to this strength.

The sedimentation test on the *render coat* revealed a high amount of seed pods and husks, a medium amount of short straw (12 mm, ½ in) and a high amount of 12 mm (½ in) lengths of soft hair. A separate sedimentation test on the binder revealed it to be comprised of 40 per cent fine sand including some fine chalk (probably remnants of the limewash), 40 per cent silt and 20 per cent clay. The proportion of binder to additives was about 1:5–8.

Conclusions

The *base coat* of the Ivy House daub was composed of a fine, well-graded, silty soil. The proportions were approximately 5 per cent clay with equal proportions of fine silt, silt and fine sand. About 5 per cent small chalk aggregate, some hair, seeds and dung had been added but not mixed in thoroughly. The *render coat* was a fine binder of 2 parts fine sand:2 parts silt:1 part clay mixed one part to 5–8 parts of seed pods, husks, ½ in straw and ½ in lengths of soft hair. The render coat was originally coated with white limewash.

General conclusions

The five daubs analysed were generally fine, well-graded soils with low clay content of 5–10 per cent. Both the sand and silt contents were in the range 20–55 per cent. While some daubs contained crushed chalk aggregate, none contained lime binder. All the samples contained a small amount of dung, probably cow dung and possibly horse dung in some cases. All the samples had straw in about 4 in lengths added.

The finest graded mixes were the most difficult to break by hand. The coarsest daub from Walderton Cottage, was the easiest to break and the most difficult to reconstitute into a workable mix.

From the evidence found within the daubs it would seem likely that the dung used was not fresh but dried and possibly weathered.

The degree of accuracy achieved by the methods of analysis used was not high but adequate regarding the degree of advancement of the construction involved. The results are considered a good guide to the constituents of the daubs submitted.

The technique of applying and drying daub is probably just as important as using an appropriate mix. The constituents must be thoroughly mixed, a minimum of water added to achieve a firm but workable consistency and the panels must be dried out gradually. The right balance of stabilizers such as sand, aggregate and straw must be achieved. Daub panels were most likely constructed two sides at a time. The details of technique need to be investigated by the construction of new daub panels and an assessment made of their behaviour during and after drying.